Applications of Computer Vision in Automation and Robotics

Applications of Computer Vision in Automation and Robotics

Editor

Krzysztof Okarma

MDPI • Basel • Beijing • Wuhan • Barcelona • Belgrade • Manchester • Tokyo • Cluj • Tianjin

Editor
Krzysztof Okarma
West Pomeranian University of
Technology in Szczecin
Poland

Editorial Office
MDPI
St. Alban-Anlage 66
4052 Basel, Switzerland

This is a reprint of articles from the Special Issue published online in the open access journal *Applied Sciences* (ISSN 2076-3417) (available at: https://www.mdpi.com/journal/applsci/special_issues/A_Computer_Vision).

For citation purposes, cite each article independently as indicated on the article page online and as indicated below:

LastName, A.A.; LastName, B.B.; LastName, C.C. Article Title. *Journal Name* **Year**, *Article Number*, Page Range.

ISBN 978-3-03943-581-4 (Hbk)
ISBN 978-3-03943-582-1 (PDF)

Cover image courtesy of Krzysztof Okarma.

© 2020 by the authors. Articles in this book are Open Access and distributed under the Creative Commons Attribution (CC BY) license, which allows users to download, copy and build upon published articles, as long as the author and publisher are properly credited, which ensures maximum dissemination and a wider impact of our publications.
The book as a whole is distributed by MDPI under the terms and conditions of the Creative Commons license CC BY-NC-ND.

Contents

About the Editor vii

Krzysztof Okarma
Applications of Computer Vision in Automation and Robotics
Reprinted from: *Appl. Sci.* **2020**, *10*, 6783, doi:10.3390/app10196783 1

Yajun Chen, Peng He, Min Gao and Erhu Zhang
Automatic Feature Region Searching Algorithm for Image Registration in Printing Defect Inspection Systems
Reprinted from: *Appl. Sci.* **2019**, *9*, 4838, doi:10.3390/app9224838 5

Fabrizio Cutolo, Umberto Fontana, Nadia Cattari and Vincenzo Ferrari
Off-Line Camera-Based Calibration for Optical See-Through Head-Mounted Displays
Reprinted from: *Appl. Sci.* **2020**, *10*, 193, doi:10.3390/app10010193 23

Yunfan Chen and Hyunchul Shin
Pedestrian Detection at Night in Infrared Images Using an Attention-Guided Encoder-Decoder Convolutional Neural Network
Reprinted from: *Appl. Sci.* **2020**, *10*, 809, doi:10.3390/app10030809 43

Bilel Benjdira, Adel Ammar, Anis Koubaa and Kais Ouni
Data-Efficient Domain Adaptation for Semantic Segmentation of Aerial Imagery Using Generative Adversarial Networks
Reprinted from: *Appl. Sci.* **2020**, *10*, 1092, doi:10.3390/app10031092 61

Petra Đurović, Ivan Vidović and Robert Cupec
Semantic Component Association within Object Classes Based on Convex Polyhedrons
Reprinted from: *Appl. Sci.* **2020**, *10*, 2641, doi:10.3390/app10082641 85

Andrius Laucka, Darius Andriukaitis, Algimantas Valinevicius, Dangirutis Navikas, Mindaugas Zilys, Vytautas Markevicius, Dardan Klimenta, Roman Sotner and Jan Jerabek
Method for Volume of Irregular Shape Pellets Estimation Using 2D Imaging Measurement
Reprinted from: *Appl. Sci.* **2020**, *10*, 2650, doi:10.3390/app10082650 105

Ibon Merino, Jon Azpiazu, Anthony Remazeilles and Basilio Sierra
Reprinted from: *Appl. Sci.* **2020**, *10*, 3701, doi:10.3390/app10113701 125

Tadej Peršak, Branka Viltužnik, Jernej Hernavs and Simon Klančnik
Vision-Based Sorting Systems for Transparent Plastic Granulate
Reprinted from: *Appl. Sci.* **2020**, *10*, 4269, doi:10.3390/app10124269 143

Krzysztof Okarma, Jarosław Fastowicz, Piotr Lech and Vladimir Lukin
Quality Assessment of 3D Printed Surfaces Using Combined Metrics Based on Mutual Structural Similarity Approach Correlated with Subjective Aesthetic Evaluation
Reprinted from: *Appl. Sci.* **2020**, *10*, 6248, doi:10.3390/app10186248 157

About the Editor

Krzysztof Okarma (Dr Hab., Assoc. Prof.) was born in Szczecin on Mar 22, 1975, graduated from the secondary school no. 4 in Szczecin with honors in 1994 and Szczecin University of Technology (currently West Pomeranian University of Technology in Szczecin—ZUT) in 1999 with honors (electronics and telecommunication) and 2001 (computer science). Since 1999, he has worked with ZUT as an assistant, PhD student, Assistant Professor and Associate Professor. Currently he is the Head of Department of Signal Processing and Multimedia Engineering and Dean of Faculty of Electrical Engineering (since 2016). He defended the PhD thesis in electrical engineering (specialty: signal processing) in 2003 and obtained the habilitation in automation and robotics (specialty: applied computer science) in 2013. The topic of his habilitation was related to image quality assessment methods, particularly applications of combined metrics. He was the auxiliary supervisor of one PhD candidate, reviewer of one habilitation and five PhD theses (including one in Lithuania) and currently is a supervisor in four projects registered for PhD degree conferment procedures. He is an author or co-author of 2 granted patents and over 200 journal and conference papers (including 25 papers in JCR journals, 10 conference papers included in CORE database and over 50 other conference papers indexed in Web of Science or Scopus databases). His papers were cited more than 300 times according to Web of Science (h-Index 10) and more than 500 times according to Scopus (h-Index 13). He was the guest editor in two Special Issues in JCR-indexed journals and a member of scientific boards of two other JCR-indexed journals and several international conferences. Currently he is the chairman of the Board of Control in the Polish chapter of IAPR (Association for Image Processing). He has also been the supervisor of over 50 master's and engineering theses and the reviewer of over 300 papers for international journals and conferences.

Editorial

Applications of Computer Vision in Automation and Robotics

Krzysztof Okarma

Department of Signal Processing and Multimedia Engineering, West Pomeranian University of Technology in Szczecin, 70-313 Szczecin, Poland; okarma@zut.edu.pl

Received: 18 September 2020; Accepted: 25 September 2020; Published: 28 September 2020

Keywords: image analysis; machine vision; video analysis; visual inspection and diagnostics; industrial and robotic vision systems

---

Computer vision applications have become one of the most rapidly developing areas in automation and robotics, as well as in some other similar areas of science and technology, e.g., mechatronics, intelligent transport and logistics, biomedical engineering, and even in the food industry. Nevertheless, automation and robotics seems to be one of the leading areas of practical applications for recently developed artificial intelligence solutions, particularly computer and machine vision algorithms. One of the most relevant issues is the safety of the human–computer and human–machine interactions in robotics, which requires the "explainability" of algorithms, often excluding the potential application of some solutions based on deep learning, regardless of their performance in pattern recognition applications.

Considering the limited amount of training data, typical for robotics, important challenges are related to unsupervised learning, as well as no-reference image and video quality assessment methods, which may prevent the use of some distorted video frames for image analysis applied for further control of, e.g., robot motion. The use of image descriptors and features calculated for natural images captured by cameras in robotics, both in "out-hand" and "in-hand" solutions, may cause more problems in comparison to artificial images, typically used for the verification of general-purpose computer vision algorithms, leading to a so-called "reality gap".

This Special Issue on "Applications of Computer Vision in Automation and Robotics" brings together the research communities interested in computer and machine vision from various departments and universities, focusing on both automation and robotics as well as computer science.

The paper [1] is related to the problem of image registration in printing defect inspection systems and the choice of appropriate feature regions. The proposed automatic feature region searching algorithm for printed image registration utilizes contour point distribution information and edge gradient direction and may also be applied for online printing defect detection.

The next contribution [2] presents a method of camera-based calibration for optical see-through headsets used in augmented reality applications, also for consumer level systems. The proposed fast automatic offline calibration method is based on standard camera calibration and computer vision methods to estimate the projection parameters of the display model for a generic position of the camera. They are then refined using planar homography, and the validation of the proposed method has been made using a developed MATLAB application.

The analysis of infrared images for pedestrian detection at night is considered in the paper [3], where a method based on an attention-guided encoder–decoder convolutional neural network is proposed to extract discriminative multi-scale features from low-resolution and noisy infrared images. The authors have validated their method using two pedestrian video datasets—Keimyung University (KMU) and Computer Vision Center (CVC)-09—leading to noticeable improvement of precision in comparison to some other popular methods. The presented approach may also be useful for collision avoidance in autonomous vehicles as well as some types of mobile robots.

Another application of neural networks has been investigated in the paper [4], where the problem of semantic segmentation of aerial imagery is analyzed. The proposed application of Generative Adversarial Networks (GAN) architecture is based on two networks with the use of intermediate semantic labels. The verification of the proposed method has been conducted using Vaihingen and Potsdam ISPRS datasets.

Since the semantic scene analysis is also useful in real-time robotics, an interesting fast method for semantic association of the object's components has been proposed in the paper [5]. The Authors have proposed an approach based on the component association graph and a descriptor representing the geometrical arrangement of the components and have verified it using a ShapeNet 3D model database.

Another application of machine vision is considered in the paper [6], where the problem of volume estimation of irregular shape pellets is discussed. The use of granulometric analysis of 2D images proposed by the authors has been verified by measurements in a real production line. The obtained results make it possible to apply a continuous monitoring of production of pellets.

Merino et al. [7] have investigated the combination of histogram based descriptors for recognition of industrial parts. Since many industrial parts are texture-less, considering their different shapes, in view of lack of big datasets containing images of such elements, the application of handcrafted features with Support Vector Machine has been proposed, outperforming the results obtained using deep learning methods.

A prototype sorting machine for transparent plastic granulate based on machine vision and air separation technology has been presented in the penultimate paper [8]. The vision part of the system is built from an industrial camera and backlight illumination. Hence, k-Nearest Neighbors based classification has been used to determine defective transparent polycarbonate particles, making it possible to use only completely transparent material for further reuse.

Another contribution utilizing combination based approach [9] focuses on the quality assessment of 3D printed surfaces. In this paper, an effective combination of image quality metrics based on structural similarity has been proposed, significantly increasing the correlation with subjective aesthetic assessment made by human observers, in comparison to the use of elementary metrics.

As may be concluded from the above short description of each contribution, computer vision methods may be effectively applied in many tasks related to automation and robotics. Although a rapid development of deep learning methods makes it possible to increase the accuracy of many classification tasks, it requires the use of large image databases for training. Since in many automation and robotic issues, a development of such big datasets is troublesome, costly and time-consuming or even impossible in some cases, the use of handcrafted features is still justified, providing good results as shown in most of the published papers.

Some of the presented approaches, e.g., utilizing a combination of features or quality metrics, may also be adapted and applied to some alternative applications. Therefore, the Guest Editor hopes that the presented works may be inspiring for the readers, leading to further development of new methods and applications of machine vision and computer vision methods for industrial purposes.

Acknowledgments: The Guest Editor is thankful for the invaluable contributions from the authors, reviewers, and the editorial team of *Applied Sciences* journal and MDPI for their support during the preparation of this Special Issue.

Conflicts of Interest: The author declares no conflict of interest.

References

1. Chen, Y.; He, P.; Gao, M.; Zhang, E. Automatic Feature Region Searching Algorithm for Image Registration in Printing Defect Inspection Systems. *Appl. Sci.* **2019**, *9*, 4838. [CrossRef]
2. Cutolo, F.; Fontana, U.; Cattari, N.; Ferrari, V. Off-Line Camera-Based Calibration for Optical See-Through Head-Mounted Displays. *Appl. Sci.* **2020**, *10*, 193. [CrossRef]

3. Chen, Y.; Shin, H. Pedestrian Detection at Night in Infrared Images Using an Attention-Guided Encoder-Decoder Convolutional Neural Network. *Appl. Sci.* **2020**, *10*, 809. [CrossRef]
4. Benjdira, B.; Ammar, A.; Koubaa, A.; Ouni, K. Data-Efficient Domain Adaptation for Semantic Segmentation of Aerial Imagery Using Generative Adversarial Networks. *Appl. Sci.* **2020**, *10*, 1092. [CrossRef]
5. Đurović, P.; Vidović, I.; Cupec, R. Semantic Component Association within Object Classes Based on Convex Polyhedrons. *Appl. Sci.* **2020**, *10*, 2641. [CrossRef]
6. Laucka, A.; Andriukaitis, D.; Valinevicius, A.; Navikas, D.; Zilys, M.; Markevicius, V.; Klimenta, D.; Sotner, R.; Jerabek, J. Method for Volume of Irregular Shape Pellets Estimation Using 2D Imaging Measurement. *Appl. Sci.* **2020**, *10*, 2650. [CrossRef]
7. Merino, I.; Azpiazu, J.; Remazeilles, A.; Sierra, B. Histogram-Based Descriptor Subset Selection for Visual Recognition of Industrial Parts. *Appl. Sci.* **2020**, *10*, 3701. [CrossRef]
8. Peršak, T.; Viltužnik, B.; Hernavs, J.; Klančnik, S. Vision-Based Sorting Systems for Transparent Plastic Granulate. *Appl. Sci.* **2020**, *10*, 4269. [CrossRef]
9. Okarma, K.; Fastowicz, J.; Lech, P.; Lukin, V. Quality Assessment of 3D Printed Surfaces Using Combined Metrics Based on Mutual Structural Similarity Approach Correlated with Subjective Aesthetic Evaluation. *Appl. Sci.* **2020**, *10*, 6248. [CrossRef]

© 2020 by the author. Licensee MDPI, Basel, Switzerland. This article is an open access article distributed under the terms and conditions of the Creative Commons Attribution (CC BY) license (http://creativecommons.org/licenses/by/4.0/).

Article

Automatic Feature Region Searching Algorithm for Image Registration in Printing Defect Inspection Systems

Yajun Chen [1,2], Peng He [1], Min Gao [1] and Erhu Zhang [1,2,*]

1. Department of information Science, Xi'an University of Technology, Xi'an 710048, China; chenyj@xaut.edu.cn (Y.C.); 2170820005@stu.xaut.edu.cn (P.H.); zgss_gaom@thunis.com (M.G.)
2. Shanxi Provincial Key Laboratory of Printing and Packaging Engineering, Xi'an University of Technology, Xi'an 710048, China
* Correspondence: eh-zhang@xaut.edu.cn; Tel.: +86-29-8231-2435

Received: 20 October 2019; Accepted: 7 November 2019; Published: 12 November 2019

Abstract: Image registration is a key step in printing defect inspection systems based on machine vision, and its accuracy depends on the selected feature regions to a great extent. Aimed at the current problems of low efficiency and crucial errors of human vision and manual selection, this study proposes a new automatic feature region searching algorithm for printed image registration. First, all obvious shapes are extracted in the preliminary shape extraction process. Second, shape searching algorithms based on contour point distribution information and edge gradient direction, respectively, are proposed. The two algorithms are combined to put forward a relatively effective and discriminative feature region searching algorithm that can automatically detect shapes such as quasi-rectangular, oval, and so on, as feature regions. The entire image and the subregional experimental results show that the proposed method can be used to extract ideal shape regions, which can be used as characteristic shape regions for image registration in printing defect detection systems.

Keywords: machine vision; defect inspection; image registration; feature region; contour point distribution; edge gradient direction

1. Introduction

Product surface defect detection is an important application field of machine vision, and it has been widely used in various industries, including the steel industry [1], textile industry [2], semiconductor manufacturing industry [3,4], and printing industry [5,6]. For surface defect detection systems based on machine vision, one key technology is image registration. Image registration is an image processing technique that aligns two images spatially. Machine vision inspection systems ensure that each detected image captured by the camera is aligned with the standard template image in a spatial position [7]. Selecting the appropriate registration features for different detection contents is a critical in improving registration accuracy.

At present, image registration algorithms can be roughly divided into three types: pixel grayscale-, image feature-, and transform domain-based image registration algorithms [8]. The pixel grayscale-based image registration algorithm usually finds the optimal match through a grayscale similarity measure, such as the normalized cross-correlation registration algorithm [9] and the sequential similarity detection algorithm [10]. This type of algorithm is stable and usually uses full grayscale information to measure the similarity of two images. However, pixel grayscale-based registration methods are often computationally intensive and thus achieve poor real-time performance. The image feature-based registration algorithm mainly uses image features such as corner, edge, texture, and shape. This method extracts the features of two images and uses a similarity measure to determine the spatial transformation relationship [11]. The extracted feature types determine the accuracy of image registration. The existing feature extraction operators mainly include the Harris operator [12], Forstner operator [13], SIFT operator [14], SURF operator [15], FAST operator [16], PCA-SIFT [17], SAR-SIFT [18], AB-SIFT [19], and other operators. This type of algorithm entails a minimal amount of computation and achieves strong robustness. However, for images with inconspicuous features, this type of algorithm seems to be ineffective [17]. The transform domain-based registration algorithm transforms images from the spatial domain to the frequency domain and then analyzes the image in the frequency domain to determine the registration parameters. The commonly used frequency domain transform-based image registration algorithms mainly include wavelet transform registration technology [20,21], Fourier transform registration technology [22–24] and composite registration algorithm combined with the space and frequency domain [25,26]. This type of image registration method has a strong anti-noise ability, but the calculation amount is relatively large.

Pixel grayscale-based and shape feature-based image registration algorithms are widely used in the field of printing defect detection systems based on machine vision. However, as mentioned previously, the pixel grayscale-based image registration algorithm entails a large amount of calculation, achieves low real-time performance, and is greatly influenced by illumination. In addition, the registration feature regions based on pixel grayscale are difficult to establish and are not robust because the image contents of different prints vary widely. Therefore, the current study adopts the shape feature-based image registration algorithm to align captured printed images with a standard reference template image. In existing printing defect detection systems, feature regions are selected manually. However, manual selection results in inconsistent selected regions and low efficiency. These drawbacks affect the precision of registration and cause systems to fail to meet the requirements of the highly automated visual inspection of printed matter. Hence, this study proposes an automatic feature region searching algorithm for image registration without manual marking to improve the real-time results and accuracy of the registration process. In addition, an automatic method for identifying registration subregions is proposed on the basis of the region partition of printed images and a subregional-associated searching method. The proposed method can cope with the misregistration of partial areas caused by paper deformation or slight rotation. Moreover, we propose a method of how to represent a good shape and give an effective feature region searching method. To the best of our knowledge, no previous research has explored this topic.

The main contributions of this work are fourfold. The details are as follows.

(1) An automatic feature region searching algorithm based on a combination of contour point distribution information and edge gradient direction information for image registration in printing defect detection systems was proposed for the first time. Despite the real-time requirements, the proposed algorithm is not complicated, and it solves the problems in printing defect inspection systems, such as low efficiency and inconsistent standards in the manual selection of registration feature regions.

(2) We innovatively described the elements of a good shape for registration and proposed good feature shape region searching algorithms using contour and edge gradient direction information. The descriptions of good shapes are discussed in Section 2.2.1.

(3) The registration feature region searching algorithm can be implemented on the basis of the region partition of printed images. Doing so can resolve uniform paper deformation, rotation, and registration errors. In the printing process, the paper may show minimal deformation, shift, or rotation. The deformation of each part of the paper after printing varies. In image registration, the adoption of the same transformation parameter in a whole printed image easily results in the partial registration of areas and obvious errors. The method is described in Section 3.2.

(4) This study proposes not only a registration feature region determination strategy for subregions based on region partition but also a feature region searching method associated with neighborhood subregions. In the actual automatic feature region searching process, some subregions may not

have stable registration feature regions. The proposed feature region searching method associated with neighborhood subregions can address this problem.

The paper is organized as follows. Section 2 presents the methodology, including the problem description, shape feature analysis, proposed shape searching method based on contour point distribution information, shape searching method based on edge gradient direction, and the improved algorithm based on the combination of contour point distribution information and edge gradient direction information. Section 3 discusses the experimental results of the automatic searching of entire image registration feature regions and subregions. Section 4 provides the conclusions.

2. Methodology

2.1. Description of the Problem

The schematic diagram of a web printing machine is shown in Figure 1. This machine usually consists of four color units, namely, cyan, magenta, yellow, and black. The printing cylinder is continuously rotating, and the printed image of one plate is repeating [5]. Figure 2 shows a schematic diagram of the entire printed image of one plate. The registration marks on the two sides of the image are used to register the cyan, magenta, yellow, and black color units. Printing defect detection generally uses the entire printed image of a cylinder plate as the basic detection unit. The images captured by the line camera are compared with the standard reference template image, which requires image registration.

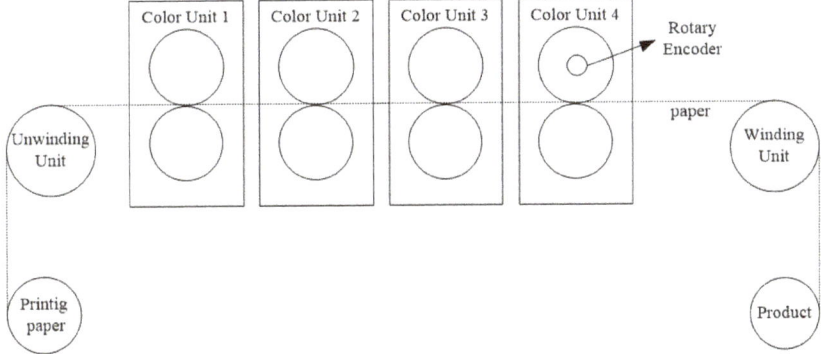

Figure 1. Schematic diagram of web printing machine [5].

Figure 2. Entire printed image of one plate and the color registration mark.

Previous printing defect inspection systems utilize the cross-line registration marks on both sides of a printed image as the registration feature. They also inspect the entire printed image of the plate as a whole. However, with the increase in printing speed, the stretching, deformation, and vibration of printing materials cause inconsistencies in the physical size of printed images. Moreover, registration with a whole plate image may cause false detection. Therefore, an increasing number of printing defect inspection systems require partition detection of entire printed images, that is, a printed image is divided into several subregions for sub-area detection. At present, the printed images of each region are not necessarily the same, and the traditional manual method of selecting a registration area is inefficient. Meanwhile, the selection of the feature regions of every sub-area by human vision is unstable and prone to errors. Therefore, an intelligent and automatic registration feature region searching method is urgently needed to achieve an effective selection of the registration feature regions of subdetection area.

Figure 3 presents the flowchart of the online printing defect detection system. We use printing defect visual inspection machine to capture the standard reference template images and other printed images in real time. During printed image defect detection, the image to be detected and the standard reference template image need to be registered initially. We propose an automatic feature region searching algorithm for image registration.

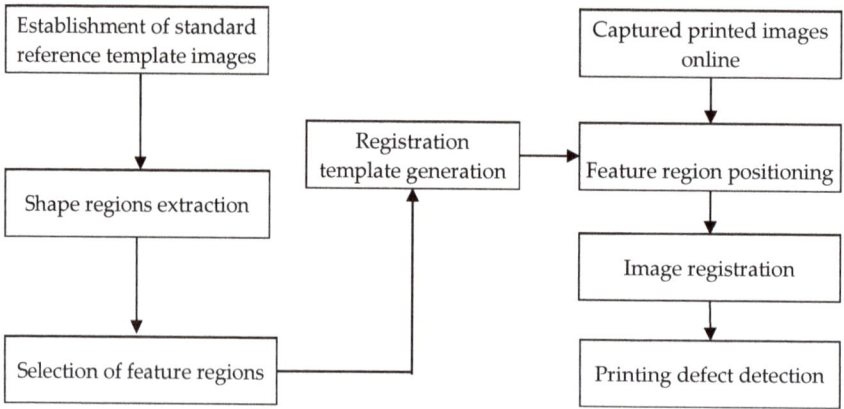

Figure 3. Flowchart of online printing defect inspection.

In this work, a region with discriminative shapes is used as the registration feature region for aligning the printed image collected online with the standard template image. A good shape region requires significant contour features and can be highly differentiated from other regions, such as geometric shapes, text, and characters, in printed images.

2.2. Shape Feature Analysis and Flow of Feature Shape Region Searching Algorithm

2.2.1. Shape Feature Analysis

As shown in Figure 4, several shapes are extracted from a printed image after the preliminary shape extraction process (Section 2.2.2). The shape in Figure 4a is mainly composed of several horizontal lines. These shapes are often included in the edges of graphics or in the complicated strokes of the characters in graphics-rich printed matter. Hence, this type of registration region shape is likely to cause a mismatch. The shape in Figure 4b belongs to a segment of a barcode. The shape feature is not obvious, and it is surrounded by many similar shapes, which could cause a mismatch. The shape in Figure 4c consists of the numeral 2 and a horizontal line. The contour of the numeral 2 is relatively regular and highly recognizable and may thus be an ideal shape feature. By contrast, the horizontal line contains little characteristic information and is almost meaningless for shape matching. In addition, it increases time consumption and is consequently undesirable. Other shapes are formed by the boundaries of the two patterns. The shape shown in Figure 4d is common in texture-rich printed matter; it is irregular and can easily cause a mismatch. On the contrary, the shape region shown in Figure 4e is a Chinese character "period" with obvious features, regular contours, and high degree of recognition feature. Thus, it is a good shape region for image registration during printed image defect detection.

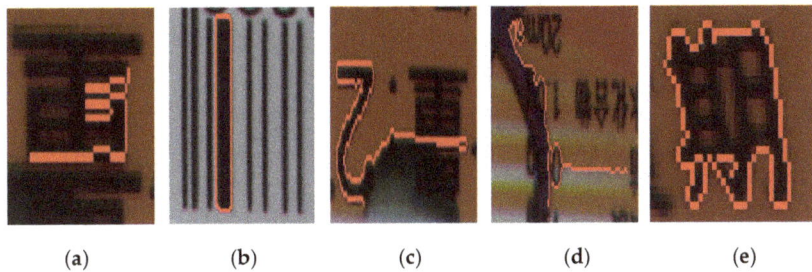

Figure 4. Several shape types from printed packaging. (**a**) Shape 1; (**b**) Shape 2; (**c**) Shape 3; (**d**) Shape 4; (**e**) Shape 5.

No research or strict definition describes the ideal shape for registration in printing defect inspection. Considering the requirements of the automatic searching of registration feature regions in online printing defect inspection, we innovatively propose a description of an ideal shape. It preferably contains the following three elements:

(1) The ideal shape should be a completely closed contour, that is, the contour points should be distributed in all direction bins.
(2) The closed shape contour should include approximately vertical and horizontal line points, that is, the gradient directions of 0°, 90°, 180°, and 270° have abundant edge points.
(3) Aside from the vertical and horizontal edge points, a good shape contour should have rich edge gradient information. In addition to the edge points in the horizontal and vertical directions, the shape contour should have many changes in contour edge direction and shape. The distribution of edge points in each gradient direction should be uniform.

The analysis shows that the feature region for printed image registration requires a completely closed contour shape, a rich contour edge gradient direction, horizontal and vertical lines, and so on. Therefore, we conclude that regions containing any one shape, such as quasi-rectangle, ellipse-like shapes, or a shape with numerous changes in edge direction, can be used as a feature region for image registration. This feature region should be easy to identify and exhibit strong robustness. This study proposes an automatic shape feature region detection algorithm that can detect quasi-rectangle, ellipse-like shapes, etc. The results should address the inaccurate and time-consuming problems of the manual selection of registration feature regions.

2.2.2. Flow of Proposed Feature Shape Region Searching Algorithm for Image Registration

The proposed feature shape region searching algorithm is shown in Figure 5. Here, "Capture printed images online" means that the printed matter for defect detection is captured on an actual printing press in real time. The proposed algorithm mainly includes four main steps.

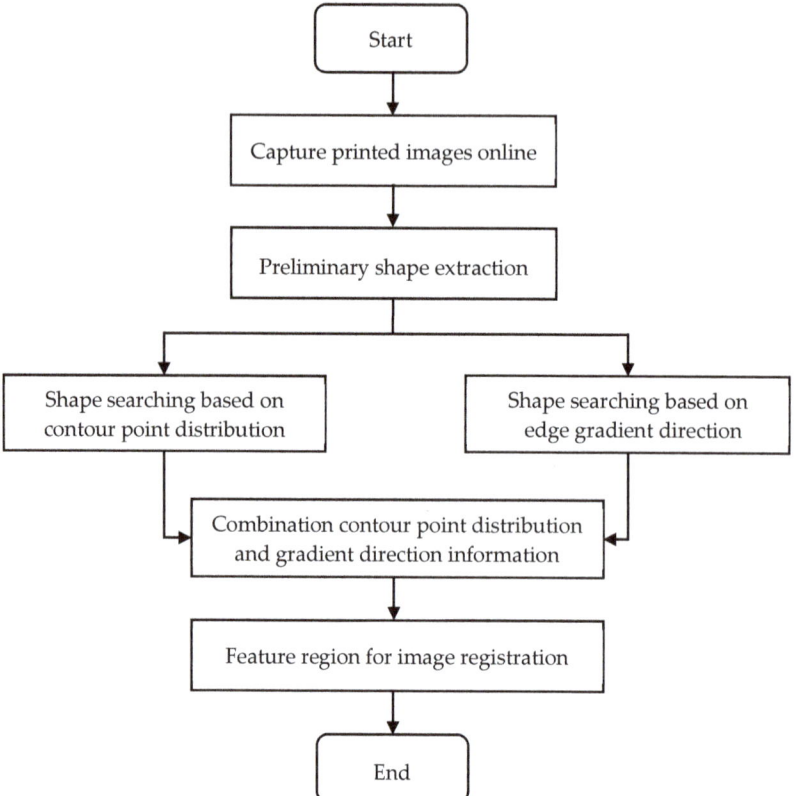

Figure 5. Flow chart of the proposed feature region searching algorithm.

(1) Prior to the implementation of the good shape searching algorithm, all shapes in the printed image should be extracted. The step is called the preliminary shape extraction. In this step, all the shapes are preprocessed, and the shape with the appropriate size is selected. The shape regions that are too small or too large are removed because excessively small shape feature affects the accuracy of image registration and an oversized shape greatly reduces the speed of the process. During the preliminary shape extraction process, we implement a series of processes on the printed images captured online. First, the adaptive segmentation of the captured image is performed, and a connected region analysis is conducted to remove the regions that are excessively small and large. Second, the shape extraction method similar to Canny edge detection is performed, and a high and low threshold idea similar to the hysteresis threshold method is used to exclude the partially inconspicuous edge contour shape. At the same time, the initially extracted shapes are recorded, and each shape is given a label number for the subsequent steps of further searching for a good shape region.

(2) The shape feature region is searched on the basis of the contour point distribution information (Section 2.3). In this step, the shape, including the edge points in several direction bins, is retained, and the shape contours that do not satisfy the judgment condition are eliminated.

(3) The shape feature region is searched on the basis of the histogram information of the edge gradient direction (Section 2.4). In this step, the shape that includes several contour edge points in four main gradient directions and contour edge points that are evenly distributed in other

gradient directions is retained. The shape contours that do not satisfy the judgment condition are eliminated.

(4) The contour point distribution information and the edge gradient histogram information are combined to propose an improved automatic feature region searching algorithm for image registration in printing defect inspection systems. The detailed description is provided in Section 2.5.

2.3. Shape searching Algorithm Based on Contour Point Distribution Information

2.3.1. Algorithm Description

The outline of a shape is composed of points, and the positional relationship of the points constitutes different shapes. The positional relationship of the shape contour points can be used to describe a shape [27]. The specific steps of the proposed algorithm are as follows.

First, the centroid position of the shape is calculated and taken as the pole to establish a polar coordinate system.

Second, the azimuth direction is divided into 12 intervals, and the number of contour points falling into each direction bin is counted, as shown in Figure 6. In this figure, point O is the centroid position, and point P is a point on the shape contour falling within the first direction bin.

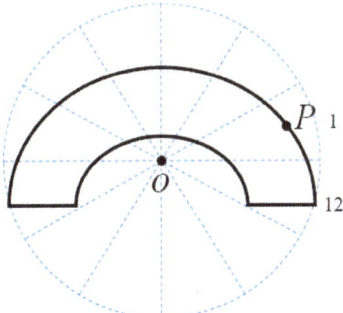

Figure 6. Shape and its contour point distribution information.

Third, we count the number of contour points in each direction bin with the assumption that the number of contour points in each direction bin is stored in a variable named *Direct*, where *Direct* $= (d_1, d_2, d_3, \ldots, d_{12})$. The contour points of a regular shape are usually evenly distributed over all directions. Therefore, we propose the following judgement conditions as Formula (1) to determine whether or not a shape is regular:

$$\begin{cases} N_d > N \\ d_{min} < S_d = \text{Deviation}(\textbf{\textit{Direct}}) < d_{max} \end{cases} \tag{1}$$

where N_d is the number of direction bins covered by all shape contour points. The operator Deviation($\cdot$) is used to calculate the normalized standard deviation of the number of contour points in each direction bin. Take Figure 3 as an example. The number of direction bins covered by the contour points of the shape is 8, N is a threshold value that defines the minimum number of direction bins covered by the contour points, and S_d is the normalized standard deviation of the number of contour points in each direction bin. The normalized standard deviation is calculated with the standard deviation of the number of contour points in each direction bin divided by the largest standard deviation of all direction bins to adapt to the contour shapes of different sizes. d_{min} and d_{max}, respectively, denote the low threshold and high threshold. The thresholds define the allowable range of the normalized standard deviation of the number of contour points in each direction bin; the more uniform the distribution direction is, the lower the standard deviation will be, resulting in a high upper-limit threshold. Note that the standard deviation of the circle contour is considerably low, that is, it is close to 0. As the circular registration region cannot determine the rotation angle of the image, the circle shape is not suitable as an image registration region. Therefore, we must define the minimum value of the standard deviation. In the actual test, the values of the above parameters are $N = 8$, $d_{min} = 0.01$, $d_{max} = 0.07$. The maximum value of n can be 11.

2.3.2. Experimental Results and Analysis

As shown in Figure 7a,b, the red shape region is the result of the shape search algorithm based on the contour point distribution information, as proposed in Section 2.2.1. Figure 7 shows that the algorithm selects the regular shape region and eliminates the similar shapes in Figure 4c,d. However, the effect of the elimination is poor, as shown in Figure 4a,b.

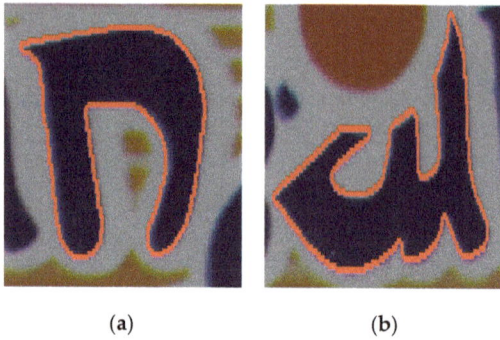

(a)   (b)

Figure 7. Experimental results of shape searching algorithm based on contour point distribution information. (a) Shape 1; (b) Shape 2.

The analysis reveals that the method takes the shape centroid as the reference point and describes the shape according to the distribution of the other points around the centroid. Although the method can judge the regularity of the shape as a whole, it cannot reflect the shape feature well. The shape shown in Figure 2a,b presents a regular contour, but the characteristics are not obvious. Therefore, the proposed method is not suitable for such cases.

2.4. Shape Searching Algorithm Based on Edge Gradient Direction

The shape of the image is outlined by the edge, and the edge is the location of gray level changes. The edge also corresponds to the boundary between the foreground and the background. Therefore, the edge gradient direction information can also reflect the shape of the image to some extent [28,29]. In general, the regular contour and obvious shape have a regular histogram of the edge gradient direction. Otherwise, the histogram of the edge gradient direction appears random. Therefore, the shape can be described on the basis of the histogram of the edge gradient direction, which is a suitable shape description method. The calculation steps of the histogram of the edge gradient direction are as follows.

First, the edge gradient of the shape is calculated using the Sobel gradient operator shown in Formulas (2) and (3), where w_1 is the horizontal gradient operator and w_2 is the vertical gradient operator:

$$w_1 = \begin{bmatrix} -1 & 0 & 1 \\ -2 & 0 & 2 \\ -1 & 0 & 1 \end{bmatrix} \qquad (2)$$

$$w_2 = \begin{bmatrix} -1 & -2 & -1 \\ 0 & 0 & 0 \\ 1 & 2 & 1 \end{bmatrix} \quad (3)$$

Second, with the assumption that the image corresponding to the shape region is $I(i,j)$, the gradient along the x direction is its horizontal gradient denoted by $G_x(i,j)$, which is given in Formula (4). Here, i, j, respectively, denote the horizontal and vertical position of the convolution calculation center at the time of Sobel extraction. The calculation is performed on the image corresponding to the minimum circumscribed rectangle of each shape region.

$$G_x(i,j) = \begin{array}{l} \{I(i-1,j+1) + 2*I(i,j+1) + I(i+1,j+1) - I(i-1,j-1) - 2*I(i,j-1) \\ -I(i+1,j-1)\},\ 0 < i < m,\ 0 < j < n \end{array} \quad (4)$$

Third, the gradient along the y direction, i.e., the vertical gradient is denoted by $Gy(i,j)$, which is given in Formula (5):

$$G_y(i,j) = \begin{array}{l} \{I(i+1,j-1) + 2*I(i+1,j) + I(i+1,j+1) - I(i-1,j-1) - 2*I(i-1,j) \\ -I(i+1,j+1)\},\ 0 < i < m,\ 0 < j < n \end{array} \quad (5)$$

Therefore, the gradient direction is given in Formula (6):

$$\theta = \arctan\left(\frac{G_y(i,j)}{G_x(i,j)}\right)\ 0 \le \theta < 2\pi \quad (6)$$

where angle θ is quantized every 5° as a direction. Thus, the interval between 0 and 2π is divided into 72 direction bins. The histogram of the edge gradient direction can be obtained by counting the number of edge points in each direction. Figure 8 illustrates the histogram of the edge gradient direction, in which every 5° is a column.

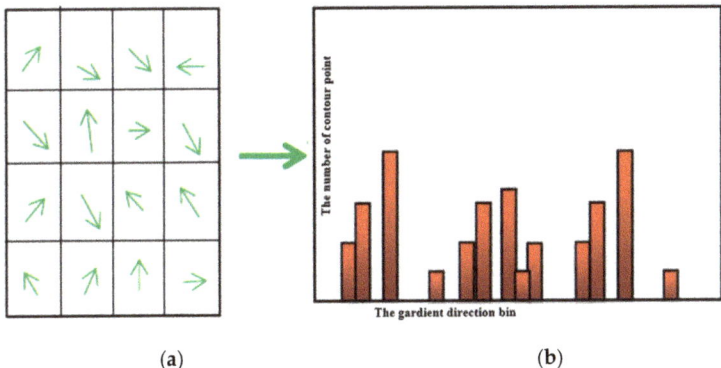

(a) (b)

Figure 8. Histogram of edge gradient direction for the shape region. (a) Schematic diagram of gradient direction. (b) Histogram of gradient direction of one shape region

Variables L and H_g represent the edge perimeter of a shape and the histogram of the edge gradient direction for storing the number of edge points in each direction. These two variables are given in Formula (7). Parameter h_i represents the number of edge points in the i-th gradient direction of each shape region.

$$\begin{cases} H_g = (h_1, h_2, \ldots, h_{72}) \\ \sum_{i=1}^{72} h_i = L \end{cases} \tag{7}$$

As shown in Figure 9, several differently shaped regions (marked as green portions) and the corresponding histograms of their edge gradient directions are given. The shape shown in Figure 9a is particularly close to a rectangle. Figure 9b shows that the edge gradient direction of the shape is mainly concentrated in the horizontal and vertical directions, that is, $\pi/2$, π, $3\pi/2$, and 2π. The four directions respectively correspond to 18, 36, 54, and 72 gradient direction bins. The histogram of the edge gradient direction distribution resembles a cross shape and a quasi-cross shape.

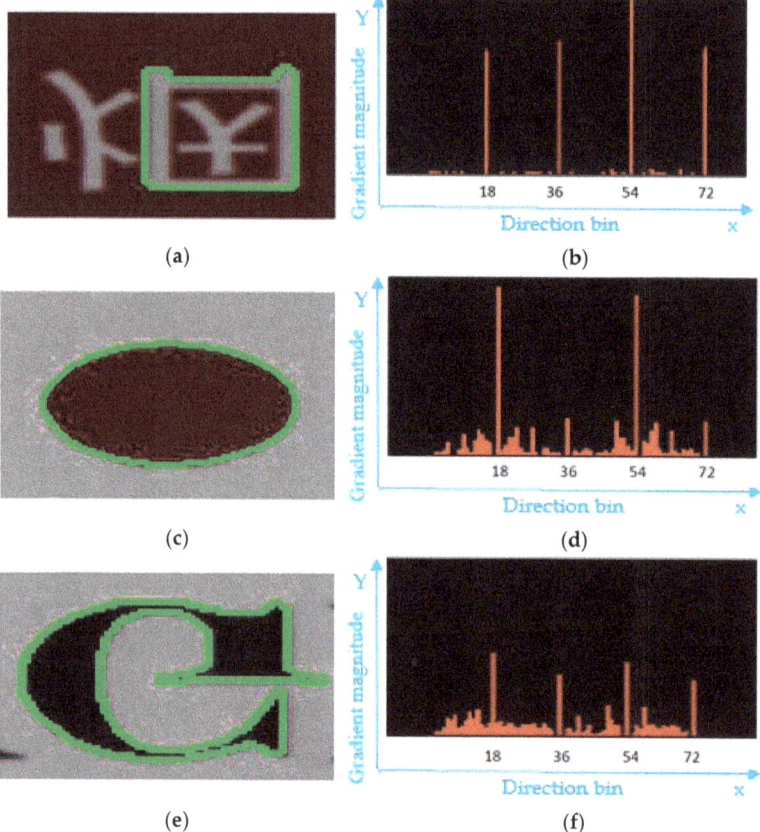

Figure 9. Experiment results of the method based on the histogram of the edge gradient direction. (**a**) Approximate rectangular shape; (**b**) shape's histogram of edge gradient direction; (**c**) elliptical shape; (**d**) shape's histogram of edge gradient direction; (**e**) good shape; (**f**) shape's histogram of edge gradient direction histogram.

Figure 9c shows an elliptical shape. The number of edge points in the vicinity of the two gradient directions ($\pi/2$, $3\pi/2$) is large, that is, the numbers of edge points in the 18th and 54th direction or nearby sections are large. The numbers of edge points in the 36th and 72nd directions or nearby sections are also large, and the numbers of edge points in the other gradient directions are relatively close. In general, the histogram distribution of the edge gradient direction is relatively uniform, as shown in Figure 9d, and is similar to the histogram distribution of the edge gradient direction of the quasi-elliptical shape. The values of the Y axis in Figure 9b,d,f are of a gradient magnitude, that is, the number of contour point falling into the direction bin.

In addition, many shapes exhibit the characteristics of the two types of regions described previously. Figure 9e shows a good shape, which has several edge points in 18, 36, 54, and 72 gradient direction bins. In addition, the numbers of edge points in the other gradient direction bins are relatively close. In general, the histogram distribution of the edge gradient direction is relatively uniform, as shown in Figure 9f.

Figure 9 shows that the aforementioned shapes have obvious features and regular contours and are thus suitable as registration feature regions. Hence, we propose to use the histogram of the edge gradient direction to detect the shape regions with rectangle, cross shape, quasi-rectangle, and quasi-cross shape types as the image registration feature regions. The detection method is defined by Formula (8):

$$h_i > T_m, \ i = 18, 36, 54, 72 \tag{8}$$

where T_m is a threshold value used to define the number of edge points in the four gradient direction bins.

Alternatively, we can select the shape region in which the overall distribution of the edge points in the gradient direction bins is relatively uniform as a feature region for image registration. The judgment conditions are as follows:

$$\begin{cases} N_g > T_n \\ S_g = \text{Deviation}(H_g) < T_g \end{cases} \tag{9}$$

where N_g is the number of directions covered by the contour points; T_n is the threshold of the number of directions, which limits the total number of directions in the lowest gradient and ensures that the shape has obvious features; S_g is the normalized standard deviation of the number of edge points in each gradient direction bin and reflects the uniformity of the distribution of edge points in each gradient direction. The size of each shape is different from the inconsistent number of edge points. Therefore, the standard deviation of the number of edge points distributed in each gradient direction bin must be normalized to ensure its versatility for differently shaped regions. T_g is a threshold that limits the maximum standard deviation of the edge points distributed over each gradient direction bin. If the condition is satisfied, then the overall distribution of the edge points of the shape in the gradient direction bin is considered to be uniform.

In the actual test, the values of the above threshold parameters are as follows: $T_m = 32$, $T_n = 25$, $T_g = 0.035$. The proposed method can extract shapes with rich features and regular contours. For example, the poor shapes of types (a) to (b) in Figure 4 show a good exclusion effect. The reason is that the edge gradient direction of the shape (a) in Figure 4 is mainly concentrated in the horizontal direction, whereas the edge gradient direction of the shape in Figure 4b is mainly concentrated in the vertical direction, and it has a single gradient direction. However, this method is also capable of extracting the shape regions of types (c) to (d) shown in Figure 4. The reason is that the edge gradients of the types (c) to (d) regions are rich and the shape is irregular. Although the basic principle of this method is based on changes in the edge gradient, it does not consider the shape as a whole. Consequently, the method cannot exclude the types (c) to (d) regions shown in Figure 4. In view of the above reasons, we propose a new feature region search method that combines the contour point distribution information and edge gradient direction.

2.5. Shape Search Algorithm Based on Combination of Contour Point Distribution and Edge Gradient Direction

The previous analysis reveals that the histogram of the edge gradient direction can effectively represent the shape feature of a region, which belongs to the global feature. The contour point distribution information can reflect the uniformity of the contour point distribution with respect to the centroid, that is, the regularity of the shape contour. The contour point distribution information belongs to local features. Therefore, a new shape searching algorithm based on the combination of the two criteria is proposed.

For a shape region with a rectangular shape, a cross shape, a quasi-rectangular shape, and a quasi-cross shape, the condition shown in Formula (10) should be satisfied, with consideration for the regularity of the shape contour. Then, the shape region can be used as the feature region for printing image registration.

$$\begin{cases} h_i > T_m, i = 18, 36, 54, 72 \\ d_{min} < S_d = \text{Deviation}(\textbf{\textit{Direct}}) < d_{max} \end{cases} \quad (10)$$

For the shape region in which the overall distribution is relatively uniform, the determination condition shown by Formula (11) is employed:

$$\begin{cases} N_g > T_n \\ S_g = \text{Deviation}(\textbf{\textit{H}}_g) < T_g \\ d_{min} < S_d = \text{Deviation}(\textbf{\textit{Direct}}) < d_{max} \end{cases}, \quad (11)$$

where the definition of each parameter is the same as the above equations.

3. Experimental Results and Analysis

3.1. Results of Automatic Feature Region Searching for Image Registration

The experimental results show that the proposed shape searching algorithm based on the combination of contour point distribution information and edge gradient direction can select the good shape in the image stably and quickly. Figure 10 provides an example of the result of a printed image. Figure 10a shows all shapes extracted after edge detection and the connected component analysis. Figure 10b shows the good shapes searched. Figure 10c presents a partially enlarged view.

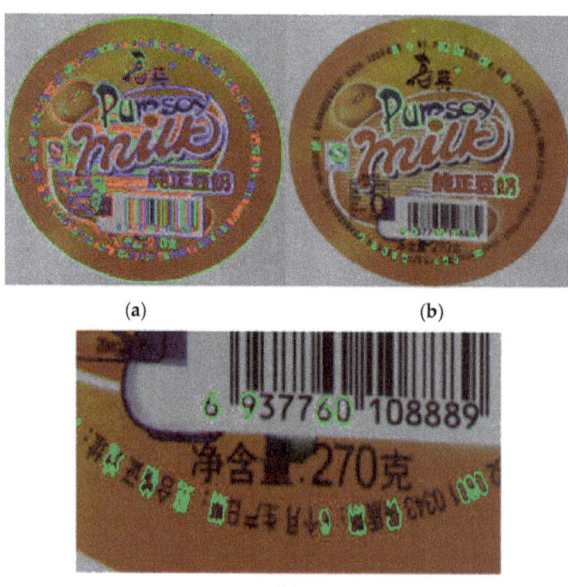

Figure 10. First set of searching results for good shapes. (**a**) All shapes detected; (**b**) good shapes searched; (**c**) partially enlarged image of Figure (**b**).

In addition to the results in Figure 10, the results of the test on the automatic registration feature region searching algorithm for other printed packaging images are given. The red contours in Figures 11a and 12a present all the shapes detected by the preliminary shape extraction method, and the red contours in Figures 11b and 12b represent good shapes searched. Figures 11c and 12c represent, respectively, the partially enlarged image of Figures 11b and 12b.

Figure 11. Second set of searching results for good shapes. (**a**) All shapes detected; (**b**) good shapes searched; (**c**) partially enlarged image of Figure (**b**).

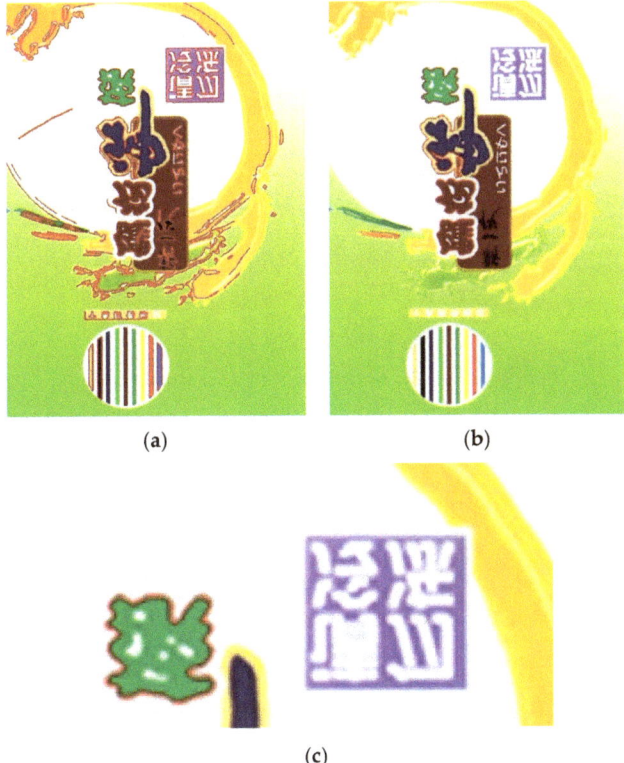

Figure 12. Third set of searching results for good shapes. (**a**) All shapes detected; (**b**) good shapes searched; (**c**) partially enlarged image of Figure (**b**).

3.2. Results of Automatic Feature Region Searching for Subregion Image Registration

A printed image is usually divided into many detection areas to overcome the influence of paper deformation during printing defect inspection. Moreover, a registration feature region is determined for each detection area. If more than one feature region is searched in a detection area, then the feature region near the center of the detection area is generally taken as the feature region for the registration of the area, and the others are discarded. If no feature region is searched in one detection area, then the neighboring feature region is taken as the registration feature region to align with the detection area. This approach is called the subregion-associated feature region searching method.

The steps of the detection zone partition and extraction of registration feature region are as follows.

(1) The detection zone is divided automatically according to the size of the printed image, which is mainly based on the appropriate number of partition lines and columns. One registration feature shape region is selected in each subregion of the large printed image.

(2) As described above, the setting of the subregion detection is no longer done by manually drawing a rectangular box. It is instead performed by an automatic division method, which can conveniently increase the number of detection areas on the detected sample image and greatly improve the automation of the modeling of the standard template reference image. As shown in Figure 13, the printed images adopt a 3 × 2 partition mode.

Figure 13. Results of subregion searching of shape feature region for image registration.

(3) The index of the detection of subregions and feature shape regions adopts the method of automatic nearby index, that is, each detected subregion selects the feature region closest to its own center point as its own index of the registration shape feature region.

(4) If the feature region is not searched in the current detected subregion, then the method of searching for the feature region associated with the adjacent subregion is adopted. That is, the feature region closest to the current subregion among all neighboring regions is selected as the registration feature region.

Figure 13 shows the subregion processing result of Figure 10. The printed image size is 1604 × 888, which is divided into six detection areas. As shown in Figure 13, each detection area is automatically searched for a suitable feature region, i.e., the green region in Figure 13.

3.3. Shape Search Parameter

After many experiments, the required parameters in the search strategy are determined. The value of each parameter is suitable when searching for shapes (Table 1). Therefore, in the actual search strategy, the parameter in Table 1 is taken as the initial value of the parameter if the condition is not found under this condition. The appropriate parameter limits of the shape can be further relaxed.

Table 1. Selected parameter values for shape search.

T_n	T_g	d_{min}	d_{max}	T_m
25	0.035	0.01	0.07	32

In the experiment, the time required to search for the best shape region in the printed image with size of 1604 × 888 pixel is 4 s. In an actual printing defect detection system, searching the feature region in the detected reference template image is set offline. Therefore, this speed of 4 s can meet the needs of the printing defect detection industry.

4. Conclusions

Automatic methods for searching feature regions in printed image registration are studied in this work, and the existing problems of the manual marker feature regions are addressed. The automatic feature region searching method for image registration without manual marking is proposed. First, the characteristics of different shapes are analyzed. The characteristics of feature region shapes for image registration are proposed. Second, the contour point distribution information and the edge gradient direction are combined. The decision conditions of the corresponding shape searching algorithm are given. The experimental results show that the proposed method can extract the ideal shape, such as a quasi-rectangle region and a quasi-ellipse region, and use it as the image registration feature region in a printing defect detection system.

The contributions of the proposed automatic feature region searching algorithms are as follows. First, despite the real-time requirements of printing defect inspection systems, the proposed algorithm for the automatic searching of registration feature regions is not particularly complicated and solves the problem in printing defect online detection systems. It also solves the problem of automatically, efficiently, and accurately establishing the registration feature regions in printing defect detection systems. The method can establish a detection standard reference template image automatically and quickly during online printing defect detection. The problem of low efficiency and unreliability of registration regions in manual searching is overcome by the proposed method, which provides an effective solution for image registration feature region searching in printing defect detection systems. Second, the description of a good shape region for image registration for printing defect inspection is proposed. Lastly, the proposed feature region extraction method for subregion registration and the subregion-associated feature region searching method solve the problem of local registration errors for the registration of entire printed images caused by the deformation and rotation of printed images. The proposed method is successfully applied to an actual online printing defect detection system for a certain enterprise.

Author Contributions: Y.C. guided the experiments and wrote the manuscript. P.H. carried out the measurements and analyzed the experimental data. M.G. designed the automatic feature region searching algorithms and wrote the program code. E.Z. conceived this study and proposed some valuable suggestions.

Funding: This work is supported by the National Key R&D Program of China under Grant No. 2018YFD0700400-2018YFD0700403, the National Natural Science Foundation of China under Grant No. 61671374, the Key Research and Development Program of Shaanxi under Grant No. 2019GY-080, the Key Program of Natural Science Foundation of Shaanxi Province of China under Grant No. 2017JZ020, the Project of Xi'an University of Technology of China under Grant No. 108-451418006, and the PhD Stand-up Fund of Xi'an University of Technology of China under Grant No. 108-256081702.

Conflicts of Interest: The authors declare no conflict of interest.

References

1. Ren, R.; Huang, T.; Chen, T.K. A generic deep-learning-based approach for automated surface inspection. *IEEE Trans. Cybern.* **2018**, *48*, 929–940. [CrossRef] [PubMed]
2. Kang, X.; Zhang, E. A universal defect detection approach for various types of fabrics based on the Elo-rating algorithm of the integral image. *Text. Res. J.* **2019**, *89*, 4766–4793. [CrossRef]
3. Vilas, H.G.; Yogesh, V.H.; Vijander, S. An efficient similarity measure approach for PCB surface defect detection. *Pattern Anal. Appl.* **2018**, *21*, 277–289.
4. Haddad, B.M.; Lina, S.Y.; Karam, L.J.; Ye, J.; Patel, N.S.; Braun, M.W. Multifeature, sparse-based approach for defects detection and classification in semiconductor units. *IEEE Trans. Autom. Sci. Eng.* **2018**, *15*, 145–159. [CrossRef]
5. Zhang, E.; Chen, Y.; Gao, M.; Duan, J.; Jing, C. Automatic defect detection for web offset printing based on machine vision. *Appl. Sci.* **2019**, *9*, 3598. [CrossRef]
6. Zhang, E.; Zhang, Y.; Duan, J. Color Inverse Halftoning Method with the Correlation of Multi-Color Com-ponents Based on Extreme Learning Machine. *Appl. Sci.* **2019**, *9*, 841. [CrossRef]
7. Cao, G.; Ruan, S.; Peng, Y.; Huang, S.; Kwok, N. Large-complex-surface defect detection by hybrid gradient threshold segmentation and image registration. *IEEE Access* **2018**, *6*, 36235–36246. [CrossRef]
8. Guo, F. *Image Registration and Defect Inspection Algorithms and Their Application in Printing System*; Beijing Jiaotong University: Beijing, China, 2011.
9. Yoo, J.C.; Han, T.H. Fast normalized cross-correlation. *Circuits Syst. Signal Process.* **2009**, *28*, 819–843. [CrossRef]
10. Gonzalez, R.C.; Woods, R.E. *Digital Image Processing*, 4th ed.; Pearson: London, UK, 2017.
11. Yamazaki, S.; Ikeuchi, K.; Shinagawa, Y. Plausible image matching: Determining dense and smooth mapping between images without a prior knowledge. *Int. J. Pattern Recognit. Artif. Intell.* **2005**, *19*, 565–583. [CrossRef]
12. Harris, C.G.; Stephens, M.J. A Combined Corner and Edge Detector. In Proceedings of the 4th Alvey Vision Conference, Manchester, UK, 31 August–2 September 1988; pp. 147–151.
13. Förstner, W.; Gülch, E. A Fast Operator for Detection and Precise Location of Distinct Points, Corners and Circular Features. In Proceedings of the ISPRS Intercommission Conference on Fast Processing of Photogrammetric Data, Interlaken, Switzerland, 2–4 June 1987; pp. 281–305.
14. Lowe, D.G. Distinctive image features from scale-invariant keypoints. *Int. J. Comput. Vis.* **2004**, *60*, 91–110. [CrossRef]
15. Bay, H.; Ess, A.; Tuytelaars, T.; Gool, L.V. Speeded-up robust features (SURF). *Comput. Vis. Image Underst.* **2008**, *110*, 346–359. [CrossRef]
16. Fularz, M.; Kraft, M.; Schmidt, A.; Kasiński, A. A high-performance FPGA-based image feature detector and matcher based on the FAST and BRIEF algorithms. *Int. J. Adv. Robot. Syst.* **2015**, *12*, 1–15. [CrossRef]
17. Yan, K.; Sukthankar, R. PCA-SIFT: A More Distinctive Representation for Local Image Descriptors. In Proceedings of the 2004 IEEE Computer Society Conference on Computer Vision and Pattern Recognition, Washington, DC, USA, 27 June–2 July 2004; IEEE: Piscataway, NJ, USA, 2004; pp. 506–513.
18. Dellinger, F.; Delon, J.; Gousseau, Y.; Michel, J. SAR-SIFT: A SIFT-Like Algorithm for SAR Images. *IEEE Trans. Geosci. Remote Sens.* **2015**, *53*, 453–466. [CrossRef]
19. Sedaghat, A.; Ebadi, H. Remote Sensing Image Matching Based on Adaptive Binning SIFT Descriptor. *IEEE Trans. Geosci. Remote Sens.* **2015**, *53*, 5283–5293. [CrossRef]
20. Raza, S.; Sanchez, V.; Prince, G.; Clarkson, J.P.; Rajpoot, N.M. Registration of thermal and visible light images of diseased plants using silhouette extraction in the wavelet domain. *Pattern Recognit.* **2015**, *48*, 2119–2128. [CrossRef]
21. Xu, Z.; Su, X. Sequence image registration based on wavelet decomposition and multi-constraint improvement. *Chin. J. Sci. Instrum.* **2011**, *32*, 2261–2266.
22. Xie, H.; Hicks, N.; Keller, G.R.; Huang, H.; Kreinovich, V. An IDL/ENVI implementation of the FFT-based algorithm for automatic image registration. *Comput. Geosci.* **2003**, *29*, 1045–1055. [CrossRef]
23. Niu, H.; Chen, E.; Qi, L.; Guo, X. Image registration based on Fractional Fourier Transform. *Optik* **2015**, *126*, 3889–3893. [CrossRef]
24. Chelbi, S.; Mekhmoukh, A. Features based image registration using cross correlation and Radon transform. *Alex. Eng. J.* **2018**, *57*, 2313–2318. [CrossRef]

25. Sun, H.; Li, Z.; Sun, L.; Lang, X. Sub-pixel registration of special and frequency domains for video sequences. *Chin. Opt.* **2011**, *4*, 154–160.
26. Nasihatkon, B.; Fejne, F.; Kahl, F. Globally Optimal Rigid Intensity Based Registration: A Fast Fourier Domain Approach. In Proceedings of the 2016 IEEE Computer Society Conference on Computer Vision and Pattern Recognition, Las Vegas, NV, USA, 27–30 June 2016; IEEE: Piscataway, NJ, USA, 2016; pp. 5936–5944.
27. Madian, N.; Jayanthi, K.B.; Somasundaram, D.; Suresh, S. Identifying Centromere Position of Human Chromosome Images using Contour and Shape based Analysis. *Measurement* **2019**, *144*, 243–259. [CrossRef]
28. Zeng, X.; Ren, L. Algorithm for image retrieval based on coherence edge orientation. *Comput. Eng. Appl.* **2012**, *48*, 205–209.
29. Yuan, X.; Hao, X.; Chen, H.; Wei, X. Robust Traffic Sign Recognition Based on Color Global and Local Oriented Edge Magnitude Patterns. *IEEE Trans. Intell. Transp. Syst.* **2014**, *15*, 1466–1477. [CrossRef]

 © 2019 by the authors. Licensee MDPI, Basel, Switzerland. This article is an open access article distributed under the terms and conditions of the Creative Commons Attribution (CC BY) license (http://creativecommons.org/licenses/by/4.0/).

Article

Off-Line Camera-Based Calibration for Optical See-Through Head-Mounted Displays

Fabrizio Cutolo [1,*], Umberto Fontana [1], Nadia Cattari [2] and Vincenzo Ferrari [1]

[1] Information Engineering Department, University of Pisa, 56126 Pisa, Italy; umbertofontana93@gmail.com (U.F.); vincenzo.ferrari@unipi.it (V.F.)
[2] EndoCAS Center, Department of Translational Research and New Technologies in Medicine and Surgery, University of Pisa, 56126 Pisa, Italy; nadia.cattari@endocas.unipi.it
* Correspondence: fabrizio.cutolo@endocas.unipi.it

Received: 15 December 2019; Accepted: 22 December 2019; Published: 25 December 2019

Featured Application: We here present a completely automated and fast method to calibrate optical see-through head-mounted displays based on the use of a calibrated camera replacing the user's eye. The method can work with any type of optical see-through display and it is easy to replicate in both laboratory environments and real-world settings.

Abstract: In recent years, the entry into the market of self contained optical see-through headsets with integrated multi-sensor capabilities has led the way to innovative and technology driven augmented reality applications and has encouraged the adoption of these devices also across highly challenging medical and industrial settings. Despite this, the display calibration process of consumer level systems is still sub-optimal, particularly for those applications that require high accuracy in the spatial alignment between computer generated elements and a real-world scene. State-of-the-art manual and automated calibration procedures designed to estimate all the projection parameters are too complex for real application cases outside laboratory environments. This paper describes an off-line fast calibration procedure that only requires a camera to observe a planar pattern displayed on the see-through display. The camera that replaces the user's eye must be placed within the eye-motion-box of the see-through display. The method exploits standard camera calibration and computer vision techniques to estimate the projection parameters of the display model for a generic position of the camera. At execution time, the projection parameters can then be refined through a planar homography that encapsulates the shift and scaling effect associated with the estimated relative translation from the old camera position to the current user's eye position. Compared to classical SPAAM techniques that still rely on the human element and to other camera based calibration procedures, the proposed technique is flexible and easy to replicate in both laboratory environments and real-world settings.

Keywords: augmented reality; calibration; head mounted displays; optical see-through display; computer vision

1. Introduction

Visual augmented reality (AR) technology aims to enhance the user's view of the real world by overlaying computer generated elements on it. Currently, optical see-through (OST) head mounted displays (HMDs) are at the leading edge of the AR technology, and they have the potential to become ubiquitous eventually in different fields of applications [1–3].

Nevertheless, their profitable usage across medical and industrial settings is still hampered by the complexity of the display calibration procedures required to ensure accurate spatial alignment between a real-world scene and computer generated elements [4,5]. The display calibration procedures of consumer level OST devices are rather simplified to improve usability, and this is achieved at the expense of sub-optimal results that are not tolerable for those applications for which the accurate alignment between virtual content and perceived reality is of the utmost importance [6]. This aspect is pushing research towards the realization of standardized OST calibration procedures that are not only flexible and easy to replicate, but that can also provide reliable and accurate results.

Overall, in visual AR applications, the problem of defining the appropriate spatial location of the digital 3D content with respect to the real scene is the principal factor that provides the user with a sense of perceptual congruity [7]. This problem is particularly challenging in OST HMDs, for the solution of which, knowing of the position of the user's viewpoint(s) cannot be overlooked. On a first approximation, a single OST display can be modeled as an off-axis pinhole camera whose imaging plane corresponds to the semi-transparent virtual screen of the display and whose projection center corresponds to the nodal point of the user's eye [8].

The goal of the OST display calibration is therefore to estimate the projection parameters of the combined eye–display system that encapsulates the optical features of the display and whose values vary according to the position of the user's eye with respect to the imaging plane of the display.

Manual calibration procedures rely on user interaction to collect sets of 3D-2D correspondences by aligning, from different viewpoints, world reference points to image points shown on the HMD virtual screen [9]. These methods, particularly when aimed at estimating all the projection parameters simultaneously, are tedious, highly dependent on operator skill, time consuming, and should be repeated every time the HMD moves on the user's head.

To lessen the burden on users in terms of time and workload, the calibration process can then be broken down into two phases: a first phase in which all the eye–display projection parameters are determined by performing a sort of factory calibration, ideally in a controlled setup; a second prior-to-use phase in which the calibration is refined by adjusting just a small subset of projection parameters. This is the underlying rationale behind several simplified manual calibration procedures, also referred to as semi-automatic calibration methods [10].

Finally, interaction-free calibration methods that exploit eye tracking techniques are ideally the preferred option for those AR applications that demand for accurate virtual-to-real spatial alignment over an extended period of time [8,11]. Such is the case of OST HMDs used as an aid to high precision tasks (e.g., for surgical or industrial applications). However, and as we illustrate in more detail in the next section, none of these methods is thoroughly automatic, since they also rely upon an off-line calibration phase performed manually or through a calibrated camera that replaces the user's eye in a controlled setup.

Therefore, irrespective of the approach chosen, the off-line calibration step is paramount to minimize the human element, and for this reason, it ought to be highly accurate and reliable.

This paper presents an approach for performing accurate off-line camera based calibrations of OST HMDs. In our procedure, the projection parameters of the eye–display model are estimated by means of standard camera calibration and computer vision techniques. To this aim, the OST HMD must be mounted over a rigid and adjustable holder equipped with a camera that replaces the user's eye.

The main contributions of our work are as follows:

1. An off-line calibration procedure that is easy to replicate in both laboratory environments and real-world settings.
2. A calibration procedure that can work with any type of OST HMD: with finite and infinite focal distances, with complex or simple optical combiners, featuring inside-out or outside-in tracking mechanisms.
3. A calibration procedure that is completely automated.
4. A calibration procedure that entails a simple on-line refinement step to account for the user's eye position.
5. A detailed formulation of all the mathematical steps involved in the procedure.

The remainder of this paper is structured as follows: Section 2 surveys different two phase approaches for OST HMD calibration. Section 3 provides a general overview of our camera based OST HMD calibration procedure together with its mathematical formulation starting from the standard projection matrix of the off-axis pinhole camera model of the eye–display. Section 4 illustrates the technical implementation of the calibration procedure. Section 5 describes the experiments and discusses the results. Finally, Section 6 concludes with a summary and future work directions.

2. Related Works

Research on how to achieve correct calibration of OST displays has been conducted over many years. In 2017, Grubert et al. [7] presented a comprehensive survey of all the calibration procedures proposed up to that time. In their work, the authors provided also useful insights into the fundamentals of calibration techniques, grouping them into three main categories: manual methods, semi-automatic methods, and automatic methods. We here provide a more focused overview of the different techniques in which the calibration workflow explicitly relies on an off-line phase.

In 2002, Genc et al. [12] proposed a simplified version of the SPAAM (a two phase SPAAM) in which the calibration process was split into two phases: an off-line stage that involves the estimation of the fixed projection parameters and a second on-line stage that aims to update the existing projection matrix partially by applying a 2D image warping of the screen image that includes scaling and shift.

A vision based robust calibration (ViRC) method was proposed in 2013 [13]. The method clearly distinguishes two types of parameters: the ones estimated off-line that are associated with the device dependent parameters and that are based on an approximate position of the user's eye and those measured on-line that are instead related to the actual position of the user's eye (4 DoF user dependent parameters). These latter parameters are refined through a perspective-n-point algorithm whose inputs are: the image coordinates of a cross-hair cursor displayed on the see-through screen at different positions and the 3D coordinates of a fiducial marker that the user must visually align to such a cursor. The 3D coordinates of the marker are determined by querying a tracking camera attached to the HMD (i.e., adopting an inside-out tracking approach).

In 2004, Owen et al. [14] presented display relative calibration (DRC), a two phase camera based calibration method. In the first phase of the method, the authors replace the user's eye with a camera and use a mechanical jig to determine the projection parameters of the display system. In the second phase, the eye–display parameters are optimized on the position of the user's eye(s) by means of a SPAAM-like procedure.

In 2005, Figl et al. [15] presented a camera based method for calibrating a medical binocular OST HMD, the Varioscope^{TM} M5. The calibration system uses a precision spindle moving a mobile calibration grid via a stepping motor and a tracker probe to localize the moving grid in space with respect to the HMD (i.e., adopting an outside-in tracking approach). However, the proposed method does not have any calibration refinement step that accounts for the user's eye position.

A similar camera based procedure was presented by Gilson et al. [16] that also employs an inside-out tracking mechanism. After the calibration, the authors did not include any user centered refinement step to account for the eye position, since in their method: "the camera remains stationary by design, so we have constrained our method to minimize for only one camera pose" (page 6). Overall, the authors declared to have achieved an average virtual-to-real alignment accuracy of about two pixels for three camera positions different from the calibration one.

In 2014, Itoh and Klinker [8,17] proposed an automatic method with inside-out tracking that employed an eye tracker to measure the position of the user's eye. The relative pose between the external camera and the eye tracking camera was predetermined during an off-line calibration session. The dynamic tracking of the user's eye was used for continuously refining the eye–display projection parameters, whose initial values once again must be computed off-line through a standard SPAAM-like method or a camera based procedure.

A similar approach was proposed in 2015 by Plopski et al. [11,18]. The method refines on-line the pre-determined calibration parameters through a corneal imaging calibration (CIC) procedure that relies on corneal imaging to obtain correspondence pairs from a pre-calibrated HMD screen with its content reflection on the cornea of the user.

In 2016, Zhang et al. [19] presented an interesting two phase method in which a depth camera refined on-line the eye–display projection parameters performing the optimization of the hand registration through a genetic algorithm. Again, a first off-line phase was needed to provide a baseline for the subsequent optimization.

Finally, in 2017, Klemm et al. [20] presented an off-line automated camera based method that resulted in an optimized rendering frustum together with an arbitrary number of distortion maps determined from different viewpoints. Unfortunately, their triangulation approach only worked for OST HMDs with simple optics for which, as stated by the authors on page 57, the "pixel locations are independent from the eye position".

As anticipated above, all these methods, require an off-line calibration step performed manually or via a camera replacing the user's eye. This off-line calibration, to be effective and accurate, should be performed in a controlled setup for the following two main reasons: to reduce the errors due to human interaction and to ease and speed-up the subsequent prior-to-use calibration refinement done by the user.

In line with this, this paper presents a camera based OST calibration method capable of accurately estimating the intrinsic matrix of the eye–display pinhole model, and we provide a detailed and substantiated formulation of each step involved in the calibration procedure.

3. Methods

3.1. Rationale

In a previous work [21], we described a closed-loop calibration method specifically suited for OST displays with the focal plane at infinity. This particular optical feature reduces the complexity of the problem of estimating the projection matrix of the off-axis eye–display pinhole model, whose parameters can be modeled irrespective of the specific eye position with a procedure based on a simple homography based 2D warping of the ideal on-axis imaging plane. In this paper, we extend the scope of such a method, so that it can be applied to any kind of OST display (i.e., with finite focal distances), and we model the contribution of the eye position to the projection parameters.

3.2. Notation

The following notation is used throughout the paper. Uppercase letters denote spatial coordinate systems, such as the eye–display coordinate system E. Lowercase letters denote scalar values, such as the focal length f_u. Both 3D and 2D points are represented by homogeneous vectors. Vectors are denoted by lowercase bold letters with a superscript denoting the reference coordinate system (e.g., a 3D point in eye–display coordinates $\mathbf{v}^E$ or a 2D image point in virtual screen coordinates $\mathbf{i}^E$). Vectors can also be expressed in component form, with a bold subscript indicating the correspondence (e.g., $\mathbf{v}^E = (x_\mathbf{v}, y_\mathbf{v}, z_\mathbf{v}, 1)^E$). Matrices are denoted by uppercase typewriter letters (e.g., the intrinsic camera matrix of the eye–display $^E\mathtt{K}$).

The 6 DoF transformations from one coordinate system to another are so defined. Given two coordinate systems A and B, the transformation from A to B is defined as $^B_A\mathtt{T} = (^B_A\mathtt{R}, ^B_A\mathbf{t})$ where $^B_A\mathtt{R}$ is the rotation matrix and $^B_A\mathbf{t}$ is the translation vector. Therefore, we have:

$$\mathbf{v}^B = {}^B_A\mathtt{R}\mathbf{v}^A + {}^B_A\mathbf{t} \tag{1}$$

3.3. The Off-Axis Pinhole Camera Model of the Eye–Display

As mentioned in Section 1, the eye–display system is commonly modeled as an off-axis pinhole camera (i.e., the associated virtual rendering camera) where the nodal point of the user's eye corresponds to the center of projection and the imaging plane is the semi-transparent virtual screen. This model describes the projection transformation between the coordinates of a point in the 3D space and the associated 2D point displayed on the imaging plane of the see-through display (Figure 1).

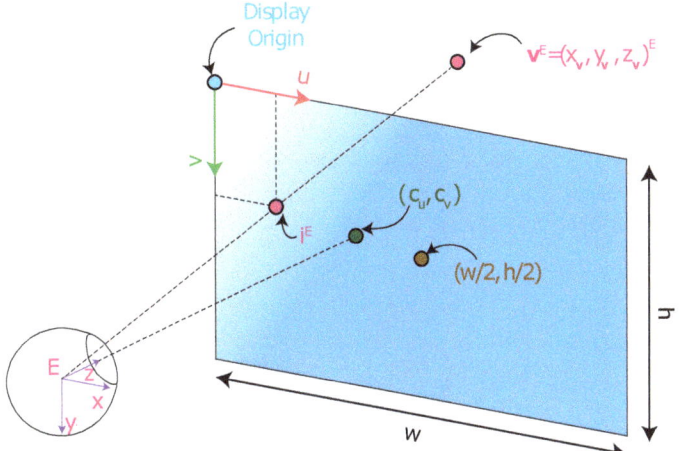

Figure 1. 3D representation of the off-axis eye–display pinhole model comprising the eye as the projection center and the see-through virtual screen as the imaging plane.

The intrinsic matrix of the off-axis eye–display model is:

$$^{\text{off}-E}K = \begin{bmatrix} f_u & 0 & c_u \\ 0 & f_v & c_v \\ 0 & 0 & 1 \end{bmatrix} \in \mathbb{R}^{3\times 3} \qquad (2)$$

where f_u and f_v are the focal lengths in pixels, and they denote the distances between the imaging plane of the display (i.e., the virtual screen) and the pinhole camera projection center (i.e., the nodal point of the user's eye); c_u and c_v are the coordinates in pixels of the principal point that corresponds to the intersection between the principal axis and the virtual screen.

Using Equation (2), the perspective projection between a 3D point in the eye–display coordinate system $\mathbf{v}^E$ and its associated 2D projection $\mathbf{i}^E$, both expressed in homogeneous coordinates, is:

$$\lambda \mathbf{i}^E = {}^{\text{off}-E}K \begin{bmatrix} I_{3\times 3} & 0_{3\times 1} \end{bmatrix} \mathbf{v}^E = {}^E P \mathbf{v}^E \qquad (3)$$

The above formulation assumes a special choice of the world coordinate system W, with $W \cong E$. In more general terms, by plugging the 6 DoF transformation from W to E into Equation (3), we obtain the projection transformation that maps world points onto the imaging plane of the eye–display $^E P \in \mathbb{R}^{3\times 4}$:

$$\lambda \mathbf{i}^E = {}^{\text{off}-E}K \begin{bmatrix} {}^E_W R & {}^E_W t \end{bmatrix} \mathbf{v}^W = {}^E P \mathbf{v}^W \qquad (4)$$

The goal of all calibration methods is to provide an accurate estimation of the matrix $^E P$, either by solving it as a whole or by determining each of the four entries of $^{\text{off}-E}K$ and each of the six DoF of $^E_W T$ individually.

3.4. Formulation of the Calibration Procedure

As mentioned in the Introduction, calibration procedures of OST displays aim to achieve a perfect spatial alignment between a 3D real-world scene and the computer generated scene perceived through the display. To this end, we need to estimate the perspective projection matrix of the off-axis rendering camera that yields a correct mapping of each 3D vertex $\mathbf{v}^W$ of the virtual object onto the imaging plane space $\mathbf{i}^E$. This 3D-2D mapping (i.e., process of rasterization) must be done in a way that all 3D vertexes along a ray that originates from the rendering camera center (i.e., the eye center E) is rendered at the same location on the virtual screen and perfectly aligned to their real counterpart. Our calibration procedure measures such a projection matrix by means of a camera that acts as a replacement of the user's eye. Hereafter, we use the term viewpoint camera to refer to such a camera.

In our formulation, as well as in the implementation of the procedure, we exploit a vision based inside-out tracking technique as done in [14,16,22]. Nonetheless, the same methodological guidelines that we here describe would also apply with an outside-in tracking technique, provided that it measures the pose of the sensor reference system, attached to the HMD, S with respect to the world scene W.

Figure 2 shows the spatial relationships between the coordinate systems associated with the tracking camera S, the internal viewpoint camera C (the user's eye replacement), and the rendering camera E.

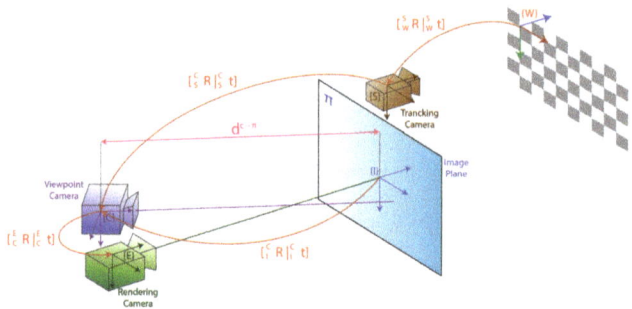

Figure 2. Illustration of the setup for the off-line calibration of the see-through display.

The first step in our calibration procedure is the estimation of the viewpoint camera projection matrix that maps world points $\mathbf{v}^W$ onto camera image points $\mathbf{i}^C$:

$$\lambda \mathbf{i}^C = {}^C K \begin{bmatrix} {}^C_S R & {}^C_S t \end{bmatrix} \begin{bmatrix} {}^S_W R & {}^S_W t \end{bmatrix} \mathbf{v}^W = \\ = {}^C K \begin{bmatrix} {}^C_W R & {}^C_W t \end{bmatrix} \mathbf{v}^W = {}^C P \mathbf{v}^W \quad (5)$$

We determine the intrinsic matrix of the viewpoint camera ${}^C K$ and the relative pose between the viewpoint camera and the tracking camera $\begin{bmatrix} {}^C_S R & {}^C_S t \end{bmatrix}$ with conventional calibration routines [23] described in more detail in Section 4.2, whereas the pose of the world scene W with respect to S is measured on-line by querying the tracking camera $\begin{bmatrix} {}^S_W R & {}^S_W t \end{bmatrix}$. We used an OpenCV checkerboard as the target object for the inside-out tracking (see Section 4.1).

Next, the key step in the procedure is the computation of the planar homography ${}^E_C H$ between the two pinhole cameras C and E that enables the image points of the viewpoint camera $\mathbf{i}^C$ to be mapped onto the imaging plane of the see-through display (i.e., the virtual screen) $\mathbf{i}^E$.

$$\lambda \mathbf{i}^E = {}^E_C H \mathbf{i}^C \quad (6)$$

where the planar homography ${}^E_C H$ can be broken down as follows [24]:

$$ {}^E_C H = {}^{on-E} K \left({}^E_C R + \frac{{}^E_C t (n^C)^T}{d^{C \to \pi}} \right) {}^C K^{-1} \quad (7)$$

In Equation (7), ^{on-E}K is the intrinsic matrix of the ideal on-axis camera model of the eye–display, whose entries are established considering the manufacturer's specifications of the OST display. Notably, we assumed that the focal lengths on both the x-axis and y-axis were equal ($f_x = f_y$), meaning the display pixels were considered as being perfectly square:

$$^{on-E}K = \begin{bmatrix} \frac{w}{2\tan(\frac{hfov}{2})} & 0 & w/2 \\ 0 & \frac{h}{2\tan(\frac{vfov}{2})} & h/2 \\ 0 & 0 & 1 \end{bmatrix} \quad (8)$$

where $\frac{hfov}{2}$ and $\frac{vfov}{2}$ are the horizontal and vertical field-of-view (FOV) of the OST display and w and h are its width and height in pixels.

The other entries of Equation (8) are:

- the rotation matrix between the imaging planes of the viewpoint camera and of the OST display $^{E}_{C}R$;
- the virtual translation vector between the viewpoint camera and the rendering camera $^{E}_{C}t$. We also label such a vector as parallax contribution, given that it models the transformation between ideal ^{on-E}K and real $^{off-E}K_{p_o}$ based on a specific eye position (subscript p_o). This contribution is described in more detail in Section 3.4.3;
- the normal unit vector to the OST display screen in the viewpoint camera reference system n^C;
- the distance between the viewpoint camera center C and the imaging plane of the OST display $d^{C \to \pi}$.

In the next two subsections, we explain how we estimate each of the above listed unknown variables.

3.4.1. Estimation of the Imaging Plane Pose with Respect to the Viewpoint Camera

We determine the position $^{I}_{C}t$ and orientation $^{I}_{C}R \equiv ^{E}_{C}R$ (i.e., the virtual pose) of the imaging plane of the OST display with respect to the viewpoint camera by computing the location of a calibration checkerboard projected onto the imaging plane of the see-through display (see Section 4).

The physical size of the checkerboard square is established by arbitrarily dictating the distance from the projection center to the imaging plane $d^{E \to \pi}$ and adopting the intrinsic parameters of the ideal on-axis eye–display model (Equation (8)), as also suggested in [25]. By observing such a checkerboard with the calibrated viewpoint camera, we are able to compute both $^{I}_{C}t$ and $^{I}_{C}R$.

Notably, the translation vector $^{I}_{C}t$ is linearly proportional to $d^{E \to \pi}$ ($^{I}_{C}t \propto d^{E \to \pi}$).

3.4.2. Estimation of the Rendering Camera Pose with Respect to the Viewpoint Camera

Given the pose of the imaging plane of the OST display with respect to a particular position of the viewpoint camera (i.e., C), we can retrieve the pose of the rendering camera relative to the viewpoint camera by considering the transformation that brings the imaging plane to the focal plane. This pose encapsulates the rotational contribution caused by the different orientations of the rendering and of the viewpoint camera $^{E}_{C}R \equiv ^{I}_{C}R$ and the parallax contribution, which is also proportional to the distance from the projection center to the imaging plane ($^{E}_{C}t \propto d^{E \to \pi}$).

Geometrically, it is easy to demonstrate the following relation, valid $\forall$ C_i position (Figure 3):

$$d^{E \to \pi} = d^{C \to \pi} + z^E_C = d^{C_i \to \pi} + z^E_{C_i} \quad (9)$$

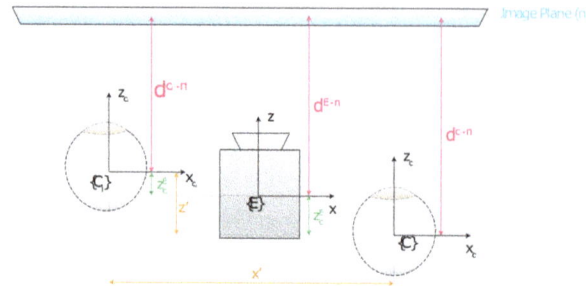

Figure 3. The x-z plane of off-axis pinhole model with the rendering camera and the eye in two different positions.

3.4.3. Meaning of the Parallax Contribution for the Off-Axis Eye–Display Pinhole Model

By plugging Equation (5) into Equation (6), we obtain:

$$\lambda i^E = {}^E_C H i^C = {}^{on-E}K \left({}^E_C R + \frac{{}^E_C t (n^C)^T}{d^{C \to \pi}} \right) [{}^C_W R \ {}^C_W t] \ v^W =$$

$$= {}^{on-E}K \left(I_{3\times 3} + \frac{{}^E_C t (n^C)^T}{d^{C \to \pi}} {}^E_C R^{-1} \right) {}^E_C R [{}^C_W R \ {}^C_W t] \ v^W = \quad (10)$$

$$= {}^{on-E}K \left(I_{3\times 3} + \frac{{}^E_C t (n^C)^T}{d^{C \to \pi}} {}^E_C R^{-1} \right) [{}^E_W R \ {}^E_W t] \ v^W$$

Since $(n^C)^T {}^E_C R^{-1} = ({}^E_C R n^C)^T = (n^E)^T = (0, 0, 1)$, which is the unit vector of the principal axis of the display, Equation (10) becomes:

$$\lambda i^E = {}^{on-E}K \left(I_{3\times 3} + \overbrace{\frac{{}^E_C t (n^E)^T}{d^{C \to \pi}}}^{^E P} \right) [{}^E_W R \ {}^E_W t] \ v^W \quad (11)$$

This can be rewritten as:

$$\lambda i^E = {}^{on-E}K \underbrace{\begin{bmatrix} 1 & 0 & x^E_C / d^{C \to \pi} \\ 0 & 1 & y^E_C / d^{C \to \pi} \\ 0 & 0 & 1 + z^E_C / d^{C \to \pi} \end{bmatrix}}_{^{off-E}K_{p_o}} [{}^E_W R \ {}^E_W t] \ v^W \quad (12)$$

${}^{off-E}K_{p_o}$ can be further processed using Equation (9) and imposing ${}^{off-E}K_{p_o}(3,3) = 1$, given that ${}^E P$ is defined up to a scale factor λ:

$$^{off-E}K_{p_o} = {}^{on-E}K \begin{bmatrix} \frac{d^{C \to \pi}}{d^{E \to \pi}} & 0 & \frac{x^E_C}{d^{E \to \pi}} \\ 0 & \frac{d^{C \to \pi}}{d^{E \to \pi}} & \frac{y^E_C}{d^{E \to \pi}} \\ 0 & 0 & 1 \end{bmatrix} \quad (13)$$

Equation (13) tells us that, to achieve an accurate spatial alignment between 3D real-world points and computer generated elements rendered by an OST display and observed by a specific viewpoint (p_o), each virtual 3D vertex v^W must be observed by an off-axis rendering camera whose projection matrix ^EP is modeled as follows:

- The extrinsic parameters of the camera are: $[^E_W R \ ^E_W t]$. Unsurprisingly, the center of projection of the virtual rendering camera is the user's eye (in our procedure, the viewpoint camera), whereas its orientation matches the display virtual screen's orientation.
- The intrinsic parameters of the off-axis camera are obtained by applying a homographic transformation to the ideal on-axis intrinsic matrix:

$$^{\text{off}-E}_{\text{on}-E}H_{p_o} = H_{p_o} = \begin{bmatrix} \frac{d^{C\to\pi}}{d^{E\to\pi}} & 0 & \frac{x^E_C}{d^{E\to\pi}} \\ 0 & \frac{d^{C\to\pi}}{d^{E\to\pi}} & \frac{y^E_C}{d^{E\to\pi}} \\ 0 & 0 & 1 \end{bmatrix} \quad (14)$$

This homography allows computing the intrinsic matrix of the real off-axis eye–display system, and it encapsulates the shift and scaling effect (i.e., the parallax contribution) due to a particular position of the user's eye.

The intrinsic matrix for a different viewpoint position p_i is:

$$^{\text{off}-E}K_{p_i} = {^{\text{on}-E}K} \begin{bmatrix} \frac{d^{C_i\to\pi}}{d^{E\to\pi}} & 0 & \frac{x^E_{C_i}}{d^{E\to\pi}} \\ 0 & \frac{d^{C_i\to\pi}}{d^{E\to\pi}} & \frac{y^E_{C_i}}{d^{E\to\pi}} \\ 0 & 0 & 1 \end{bmatrix} \quad (15)$$

Therefore, from simple algebraic manipulations, we can obtain the transformation matrix that enables us to pass from $^{\text{off}-E}K_{p_o}$ to $^{\text{off}-E}K_{p_i}$:

$$^{\text{off}-E}K_{p_i} = {^{\text{off}-E}K_{p_o}} H_{p_o}^{-1} H_{p_i} =$$

$$= {^{\text{off}-E}K_{p_o}} \begin{bmatrix} \frac{d^{C_i\to\pi}}{d^{C\to\pi}} & 0 & \frac{x^E_{C_i}-x^E_C}{d^{C\to\pi}} \\ 0 & \frac{d^{C_i\to\pi}}{d^{C\to\pi}} & \frac{y^E_{C_i}-y^E_C}{d^{C\to\pi}} \\ 0 & 0 & 1 \end{bmatrix} \quad (16)$$

From this, we derive the same relation presented in [8] and in [7]:

$$^{\text{off}-E}K_{p_i} = {^{\text{off}-E}K_{p_o}} \begin{bmatrix} 1+\frac{z'}{d^{C\to\pi}} & 0 & \frac{-x'}{d^{C\to\pi}} \\ 0 & 1+\frac{z'}{d^{C\to\pi}} & \frac{-y'}{d^{C\to\pi}} \\ 0 & 0 & 1 \end{bmatrix} \quad (17)$$

where $^{C_i}_C t = [x', y', z']$ is the translation from the old viewpoint position to the new viewpoint position and where $d^{C_i \to \pi} = d^{C \to \pi} + z'$.

3.4.4. Relation between the Optical Properties of the OST Display and Intrinsic Matrix of the Eye–Display Model

In any HMD, the role of the collimation optics is to display the microdisplay image so that it appears magnified at a comfortable viewing distance [26]. Specifically, the imaging plane of many consumer level OST HMDs is at infinity (i.e., at a very far distance), and this happens if the microdisplay is located in close proximity of the focal point of the collimation optics. In this case, the light rays coming from each pixel of the microdisplay are arranged in a parallel pattern, and the virtual image perceived by the user, to a first approximation, is not subjected to any shift or scaling effect for different positions of the user's eye within the display eye-motion-box area.

Thereby, for those OST displays, the position of the user's eye does not influence the intrinsic matrix of the off-axis eye–display pinhole camera model $^{\text{off-E}}K_p$, and this implies that the intrinsic linear parameters of the eye–display can be considered as parallax-free:

$$^{\text{off-E}}K_{p_i} = {^{\text{off-E}}K_{p_0}} \begin{bmatrix} 1 + \overbrace{\frac{z'}{d^{C \to \pi}}}^{\approx 0} & 0 & \overbrace{\frac{-x'}{d^{C \to \pi}}}^{\approx 0} \\ 0 & 1 + \overbrace{\frac{z'}{d^{C \to \pi}}}^{\approx 0} & \overbrace{\frac{-y'}{d^{C \to \pi}}}^{\approx 0} \\ 0 & 0 & 1 \end{bmatrix} \approx {^{\text{off-E}}K_{p_0}} \quad (18)$$

since $x' \backslash y' \backslash z' \ll d^{C \to \pi}$.

In this case, the on-line refinement step accounting for the user's eye position is not needed. Differently, for those OST displays with shorter focal distances (e.g., Microsoft HoloLens, Magic Leap, Epson Moverio BT-200), the light rays within the eye-motion-box of the display are not parallel. For this reason, the contribution of the position of the user's eye to the intrinsic matrix cannot be neglected, and the update contribution expressed by Equation (17) must be taken into consideration (i.e., the intrinsic matrix is not parallax-free).

4. Technical Implementation of the Procedure

4.1. Hardware Setup

Our calibration procedure is applicable to any OST HMD, with finite or infinite focal distance, based on waveguides or on large spherical or semispherical optical combiners.

We tested our method on a commercial binocular OST HMD (ARS.30 by Trivisio [27]) appropriately reworked to embody a vision based inside-out tracking mechanism. The visor was provided with dual SXGA OLED panels with $1280 \times 101,024$ resolution and with a pair of standard flat optical combiners tilted at about 45°. The two panels were controlled via HDMI. The diagonal FOV of the HMD was of 30°, which corresponded to ≈ 1.11 arcmin/pixel angular resolution. The eye-relief of the display was 30 mm, and the size of the eye-motion-box was 8×10 mm. The imaging plane of the OST display was projected at 500 mm ($d^{E \to \pi} = 500$ mm).

Following a similar approach to our previous research works [28,29], we embedded the HMD in a 3D printed plastic shell; the plastic shell was designed to house a pair of liquid crystal (LC) shutters and a pair of stereo RGB cameras (Leopard Imaging LI-OV4689) for the inside-out tracking mechanism (Figure 4). The LC panels were placed in front of the optical combiners of the HMD, and they could be electronically controlled so as to modify the transparency of the see-through display and switch from see-through state to occluded state. We used the occluded modality for the second step of the calibration, in which we projected a calibration checkerboard onto the imaging plane of the see-through display.

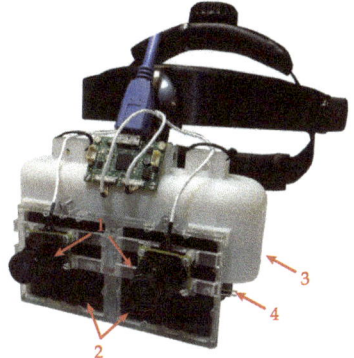

Figure 4. The custom-made hybrid video-optical see-through head mounted display used in the experimental session. 1→Pair of stereo cameras for the inside-out tracking. 2→Pair of liquid-crystal optical shutters to drive the switching mechanism between occluded to optical see-through state. 3→Plastic shell that incorporates all the components of the optical see-through visor. 4→Optical combiner of the optical see-through visor.

Figure 5 shows the experimental setup. In our procedure, we used a single camera inside-out tracking mechanism. The tracking cameras had 109° diagonal FOV, which, was associated with a 1280×720 image resolution, yielding ≈ 4.8 arcmin/pixel angular resolution. As we will discuss in Section 5, this particular feature enabled us to have a reference limit in the evaluation of the overall calibration accuracy.

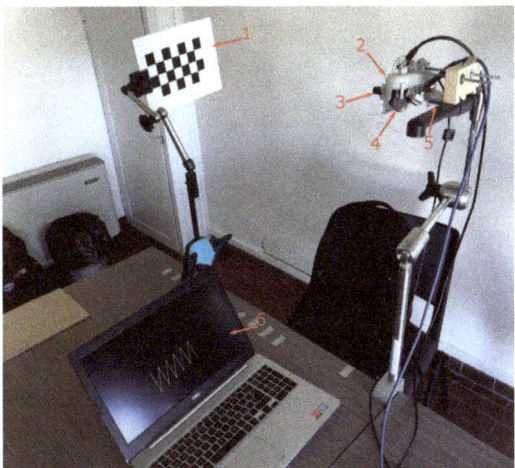

Figure 5. Experimental environment. 1→Validation checkerboard. 2→Optical see-through head mounted display. 3→Tracking camera. 4→Optical combiner. 5→3D printed mounting template with the embedded viewpoint camera. 6→Laptop where the calibration routines and the AR application run.

In our tests, we performed the calibration on the right display. Therefore, we placed the viewpoint camera behind the right display at approximately the eye-relief distance from the flat combiner (i.e., ≈ 30 mm).

The viewpoint camera was a SONY FCB-MA130, which had a 1/2.45" CMOS sensor, a 1280 × 720 resolution, and a 59° diagonal FOV, which corresponded to ≈ 2.67 arcmin/pixel.

Both the HMD and the validation checkerboard were locked to two rigid and adjustable holders. The validation checkerboard was a standard 7 × 4 OpenCV calibration checkerboard with a square size of 30 mm. The checkerboard was placed at approximately 650 mm of distance from the viewpoint camera. We attached the viewpoint camera to a 3D printed mounting template. The mounting template was equipped with fixing holes for placing the camera in eight pre-set positions radially arranged within the eye-motion-box of the see-through display (Figure 6). The template and the camera were both anchored to the translation bar of the HMD holder.

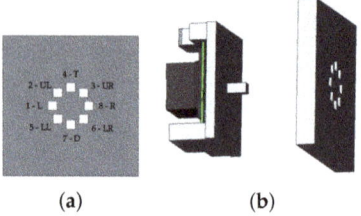

(a)   (b)

Figure 6. 3D CAD of the mounting template used for placing the viewpoint camera in eight different positions within the eye-motion-box of the see-through display. (**a**) The eight fixing holes for camera positioning. (**b**) The mounting template.

We calibrated the display for the left position (reference position), and then, we computed the intrinsic matrix of the eye–display for the remaining seven positions using Equation (17), where $d^{C\to\pi}$ is given by Equation (9) and where the translation vector $^{C_i}_C t$ is computed using two stereo calibrations between the tracking camera (fixed) and the viewpoint camera in the reference and in the new position:

$$^{C_i}_C t = [x', y', z'] = {^S_C t} - {^S_{C_i} t} \tag{19}$$

4.2. Calibration Software

As outlined in Section 3.4, the calibration workflow was broken down into four main steps, each of which was associated with a different software routine:

- Estimation of the intrinsic camera parameters of both the viewpoint camera and the tracking camera (intrinsic camera calibrations).
- Estimation of the pose of the tracking camera with respect to the viewpoint camera (heterogeneous stereo calibration).
- Estimation of the pose of the rendering camera with respect to the viewpoint camera (parallax contribution estimation).
- Estimation of the final eye–display projection matrix (eye–display matrix calculation).

The calibration workflow is depicted in Figure 7. The details of the calibration steps are presented in the next subsections.

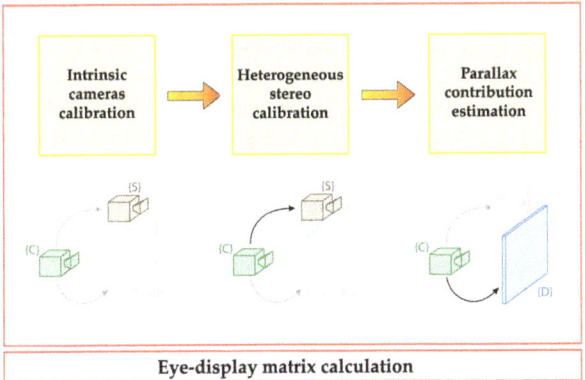

Figure 7. Conceptual schematic of the optical see-through display calibration procedure.

4.2.1. Intrinsic Camera Calibrations

The viewpoint camera and the tracking camera were calibrated with a conventional calibration technique [23] that required storing multiple camera views of a planar pattern (i.e., OpenCV checkerboard). Linear parameters (i.e., intrinsic camera matrix) and non-linearities due to camera lens distortion (i.e., distortion parameters) were computed using non-linear least-squares minimization (i.e., the Levenberg–Marquardt algorithm). This procedure was performed using the MATLAB camera calibration toolbox (R2018b MathWorks, Inc., Natick, MA, USA).

4.2.2. Heterogeneous Stereo Calibration

The relative pose between the viewpoint camera and the tracking camera was estimated through a stereo calibration routine specific for two different camera types. This algorithm was developed in C++ under the Linux Operating System (Ubuntu 16.04) and using the OpenCV API 3.3.1 [30].

4.2.3. Parallax Contribution and Eye–Display Matrix Estimation

The output of the heterogeneous stereo calibration algorithm, together with the intrinsic parameters of the two physical cameras and the manufacturer's ideal projection parameters of the display were all fed into the final calibration routine. This routine was developed in MATLAB.

As described in Section 3.4, the pose of the OST display imaging plane with respect to the viewpoint camera was determined as follows: a calibration checkerboard was projected onto the imaging plane of the occluded see-through display in full-screen modality and acquired by the viewpoint camera; the virtual pose of the imaging plane was determined through a standard perspective-n-point algorithm [31]. These final data enabled us to compute the projection parameters of the off-axis eye–display model.

5. Experiments and Results

A dedicated software application was developed in MATLAB to validate the accuracy of the calibration technique. In the application, we generated a virtual scene whose virtual viewpoint (i.e., rendering camera) was controlled according to the extrinsic and the intrinsic parameters of the eye–display model determined from the calibration. In this way, we created an AR OST visualization, whose overlay accuracy could be considered as an objective evaluation metric for the calibration procedure. We therefore elaborated the image frames of the viewpoint camera and measured the overlay error, or reprojection error, between real and virtual features.

As ground truth real features, we considered the corners of the validation checkerboard, whereas as virtual features, we used virtual circles. The overlay error was computed as the Euclidean distance (o_{2D}) between the image coordinates of the real and virtual features (Figure 8).

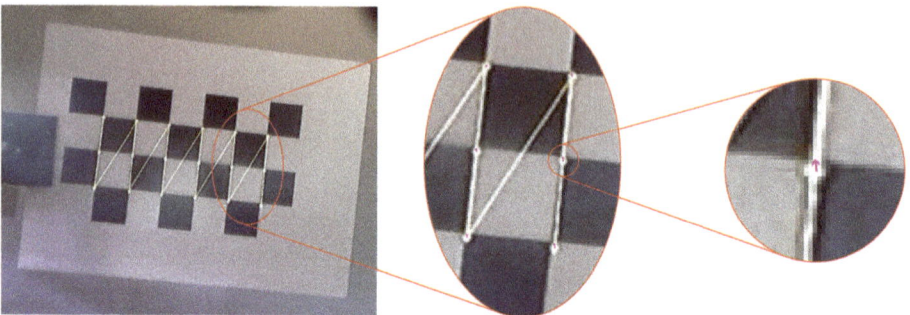

Figure 8. Image grabbed from the viewpoint camera after the calibration. The accuracy of the calibration is measured as the overlay error between the ground truth real corners of the validation checkerboard and the centers of the virtual yellow circles. The zoomed details of the image shows the on-image displacements between the real and the virtual landmarks. Red arrows indicate the direction and magnitude of overlay error in pixels.

We exploited the hybrid nature of the HMD, which can work both under OST and the video see-through modality [28], to measure the overlay error as follows:

- The image coordinates of the checkerboard corners were extracted by processing a frame of the viewpoint camera acquired with both the display and the optical shutter turned off. In order to do this, we used the MATLAB corner detection algorithm (Figure 9b).
- The image coordinates of the virtual landmarks were extracted by processing the viewpoint camera image without the real-world background, hence with both the display and the optical shutter turned on. The coordinates of the centroids of the virtual circles were retrieved using a semi-automatic centroid detection algorithm (Figure 9c).

5.1. Quantitative Evaluation of Virtual-to-Real Registration Accuracy

Quantitative results are presented in terms of the average value, standard deviation, and max value of the overlay error (o_{2D}^{err}) onto the imaging plane of the viewpoint camera over the eight positions on which the calibration was performed. Similarly, we also measured the values of the subtended angular error (α^{err}). Then, knowing the overlay error in pixels and given the distance z_C^g and the focal length of the viewpoint camera f_C, we also computed the associated absolute error in mm at the validation checkerboard plane by using the following relation: [32,33]:

$$a_{3D}^{err} \approx \frac{o_{2D}^{err} z_C^g}{f_C} \tag{20}$$

5.2. Results and Discussion

Table 1 shows the results of the eight calibrations in terms of overlay error (o_{2D}^{err}), angular error (α^{err}), and absolute error (a_{3D}^{err}). The overall mean, standard deviation, and max values of o_{2D}^{err} were 2.23, 0.96, and 5.03 px; the same values for α^{err} were 5.98, 2.58, and 13.47 arcmin and for a_{3D}^{err} 1.13, 0.49, and 2.55 mm.

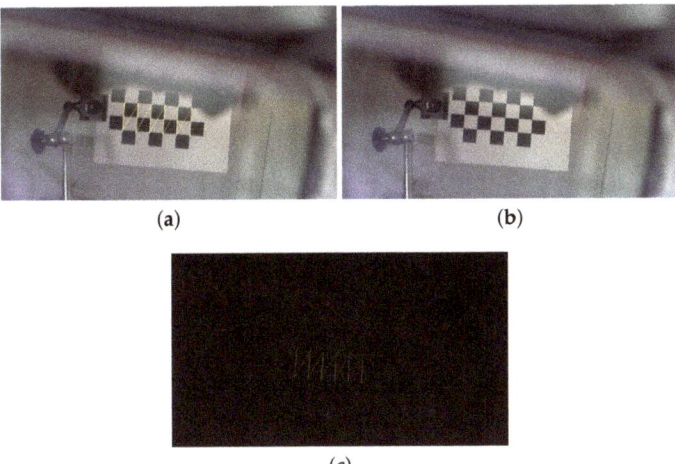

Figure 9. Viewpoint camera frames. (**a**) Camera frame of the augmented scene. (**b**) Camera frame of the real-world scene. (**c**) Camera frame of the virtual scene.

Table 1. Results of the accuracy analysis in pixels (overlay error), arcmin (angular error), and mm (absolute error).

Camera Position	Overlay Error (pixel)			Angular Error (arcmin)			Absolute Error (mm)		
	Mean	Std Dev	Max	Mean	Std Dev	Max	Mean	Std Dev	Max
1-L	1.08	0.52	1.98	2.88	1.39	5.3	0.55	0.26	1
2-UL	1.43	0.78	3.23	3.82	2.08	8.66	0.72	0.39	1.64
3-UR	2.03	0.87	3.48	5.43	2.32	9.33	1.03	0.44	1.76
4-T	2.06	0.58	3.08	5.52	1.56	8.25	1.04	0.3	1.56
5-LL	2.74	0.82	4.34	7.35	2.2	11.61	1.39	0.42	2.19
6-LR	3.29	0.6	5.03	8.82	1.6	13.47	1.67	0.3	2.55
7-D	2.61	0.66	4	6.99	1.76	10.72	1.32	0.33	2.03
8-R	2.61	0.67	4.05	6.99	1.8	10.85	1.32	0.34	2.05

The average overlay error for the reference position was comparable to that obtained by Owen et al. [14] and by Gilson et al. [16], whereas for the remaining positions, the average overlay error was comparable to that obtained in [9] and in [8]. It should be also noted that the results of the angular error are reported considering the rather low angular resolution of the viewpoint camera ($\approx$2.67 arcmin/pixel). This explained the values of the angular errors obtained. If we considered the absolute error at the distance where the validation checkerboard was placed ($\approx$650 mm), our results were comparable to those obtained with the camera based triangulation approach proposed by Klemm et al. [20]. The experiment results suggested that the calibration refinement step accounting for the user's eye position was paramount to achieve accurate results also for eye positions different from the reference one for all those OST HMDs with finite focal distances (i.e., not parallax-free OST HMDs). This was in accordance with what was suggested by Owen et al. [14] and contrary to what was hypothesized by Gilson et al. and by [34] et al. It should be also stressed that this assumption was clearly different from what was suggested in [20] (page 58), where the classification of the OST HMDs was made based on the level of complexity of the OST optical combiner: "these results show that the triangulation approach only works for OSTG with simple optics and that it is not a universal approach".

Overall, there were at least two sources of calibration inaccuracies in our method, both of which could be easily addressed in the future. The first one was not due to the calibration procedure per se, but it was due to the intrinsic tracking accuracy of the inside-out mechanism adopted in our study, which, as anticipated, was affected by the angular resolution of the tracking camera (i.e., ≈4.8 arcmin/pixel). This choice was dictated by the need to offer sufficient stereo overlap between the two tracking cameras, albeit at the expense of a reduced pixel density. It should be noted that this value of angular resolution had a detrimental effect on the calibration results as it increased the uncertainty in estimating the position of the validation checkerboard. This measure was used both for calibrating the eye–display model for the reference position of the viewpoint camera and also for computing the translation vector $^{C_i}_{e}t$. We believe that in the future, the accuracy of the off-line phase of the calibration would benefit from the selection of tracking cameras with a higher value of angular resolution, achievable by selecting cameras with a higher image resolution or with longer focus lenses.

A second source of inaccuracy arose from not considering, in our solution, the contribution of the display optical distortion. Indeed, the collimation optics of any OST HMDs, despite being essential to provide a reasonable FOV and a comfortable viewing distance for the virtual image, is a source of optical distortions such as radial and tangential distortions [35]. The same can be said for the optical combiner, particularly for waveguide based OST displays [20]. These optical distortions not only affect the shapes of the virtual objects perceived by the user, but they also distort the pattern of binocular disparities between the left and right virtual images, and therefore, they alter the stereoscopic perception of the virtual content [36]. To counter this problem, we are currently working on a solution that integrates a non-linear optimization step in the calibration algorithm. This further step can enable us to estimate the optical distortion parameters (radial and tangential). By doing so, we can correctly compensate for the optical distortions through a predistortion technique based on a 2D non-linear mapping of the virtual image on the imaging plane, as done in [25].

In order for the distortion correction, as well as the transformation (17) to be efficient and real time, a GPU based texture mapping approach (i.e., late warping) should be also adopted [37]. This mechanism would require the use of an eye tracking camera. On account of this, our calibration procedure potentially can be integrated in the low level rendering mechanism of commercial OST HMDs to encompass the linear and non-linear 2D warping of the ideal imaging plane yielded by the intrinsic optical properties of the display and by the user's eye position [38].

6. Conclusions and Future Work

In this paper, we presented a method to easily calibrate OST HMDs based on the use of a calibrated camera replacing the user's eye. The method, which could work with any type of OST HMD, only required the camera to observe a planar calibration pattern displayed on the imaging plane of the see-through display.

The method exploited standard camera calibration and computer vision techniques for estimating the projection parameters of the off-axis display model for a generic position of the viewpoint camera. In a prior-to-use second phase, the projection parameters were then refined through a planar homography that encapsulated the shift and scaling effect (i.e., the parallax contribution) associated with the relative translation from the old camera position to the current user's eye position.

To evaluate the efficacy of the calibration technique objectively, we measured the overlay error between real and virtual features observed by the viewpoint camera from different positions within the eye-motion-box of the display. Experimental results indicated that the average overlay error of our calibration was 2.23 pixel.

Our future work involves the integration of an additional step in the off-line calibration procedure devoted to the estimation of the radial and tangential distortion coefficients due to the lens system of the OST display. This calibration step can be performed through a non-linear optimization method such as the Levenberg–Marquardt algorithm.

We are also working on the integration of an eye tracking camera with our custom made OST HMD to perform user studies to evaluate the efficacy of the run-time refinement of the linear and non-linear calibration parameters and to design possible algorithm improvements that take into consideration the eye model.

Finally, the flexibility of our camera based calibration procedure can pave the way toward the profitable use of consumer level OST HMDs also in those applications that demand high accuracy in the spatial alignment between computer generated elements and the real-world scene (e.g., in surgical or high precision industrial applications).

Author Contributions: Conceptualization, F.C.; methodology, F.C., U.F., and N.C.; software, F.C., U.F., and N.C.; validation, F.C., U.F., N.C., and V.F.; formal analysis, F.C.; investigation, F.C. and U.F.; resources, F.C. and V.F.; data curation, U.F. and N.C.; writing, original draft preparation, F.C.; writing, review and editing, F.C., U.F., and N.C.; visualization, F.C.; supervision, F.C. and V.F.; project administration, V.F.; funding acquisition, V.F. All authors have read and agreed to the published version of the manuscript.

Funding: This research was supported by the HORIZON2020 Project VOSTARS (Video-Optical See Through AR surgical System), Project ID: 731974. Call: ICT-29-2016 Photonics KET2016.

Acknowledgments: The authors would like to thank Renzo D'Amato for his support in designing and assembling the AR visor.

Conflicts of Interest: The authors declare no conflict of interest. The funders had no role in the design of the study; in the collection, analyses, or interpretation of the data; in the writing of the manuscript; nor in the decision to publish the results.

Abbreviations

The following abbreviations are used in this manuscript:

AR   Augmented reality
HMD   Head mounted display
OST   Optical see-through
SPAAM Single point active alignment method
FOV   Field-of-view

References

1. Liu, H.; Auvinet, E.; Giles, J.; Rodriguez y Baena, F. Augmented Reality Based Navigation for Computer Assisted Hip Resurfacing: A Proof of Concept Study. *Ann. Biomed. Eng.* **2018**, *46*, 1595–1605. [CrossRef] [PubMed]
2. Chen, L.; Day, T.W.; Tang, W.; John, N.W. Recent Developments and Future Challenges in Medical Mixed Reality. In Proceedings of the 2017 IEEE International Symposium on Mixed and Augmented Reality (ISMAR 2017), Nantes, France, 9–13 October 2017; pp. 123–135. [CrossRef]
3. Zhou, F.; Duh, H.B.; Billinghurst, M. Trends in augmented reality tracking, interaction and display: A review of ten years of ISMAR. In Proceedings of the 2008 7th IEEE/ACM International Symposium on Mixed and Augmented Reality (ISMAR 2008), Cambridge, UK, 14–18 September 2008; pp. 193–202. [CrossRef]
4. Qian, L.; Azimi, E.; Kazanzides, P.; Navab, N. Comprehensive tracker based display calibration for holographic optical see-through head-mounted display. *arXiv* **2017**, arXiv:1703.05834.
5. Cutolo, F. Letter to the Editor on "Augmented Reality Based Navigation for Computer Assisted Hip Resurfacing: A Proof of Concept Study". *Ann. Biomed. Eng.* **2019**, *47*, 2151–2153. [CrossRef] [PubMed]
6. Jones, J.A.; Edewaard, D.; Tyrrell, R.A.; Hodges, L.F. A schematic eye for virtual environments. In Proceedings of the 2016 IEEE Symposium on 3D User Interfaces (3DUI 2016), Greenville, SC, USA, 19–20 March 2016; pp. 221–230. [CrossRef]
7. Grubert, J.; Itoh, Y.; Moser, K.; Swan, J.E. A Survey of Calibration Methods for Optical See-Through Head-Mounted Displays. *IEEE Trans. Vis. Comput. Graph.* **2018**, *24*, 2649–2662. [CrossRef] [PubMed]

8. Itoh, Y.; Klinker, G. Interaction-free calibration for optical see-through head-mounted displays based on 3D Eye localization. In Proceedings of the 2014 IEEE Symposium on 3D User Interfaces (3DUI 2014), Minneapolis, MN, USA, 19–20 March 2014; pp. 75–82. [CrossRef]
9. Tuceryan, M.; Navab, N. Single point active alignment method (SPAAM) for optical see-through HMD calibration for AR. In Proceedings of the IEEE and ACM International Symposium on Augmented Reality (ISAR 2000), Munich, Germany, 5–6 October 2000; pp. 149–158. [CrossRef]
10. Navab, N.; Zokai, S.; Genc, Y.; Coelho, E.M. An on-line evaluation system for optical see-through augmented reality. In Proceedings of the IEEE Virtual Reality 2004 (IEEE VR 2004), Chicago, IL, USA, 27–31 March 2004; pp. 245–246. [CrossRef]
11. Plopski, A.; Itoh, Y.; Nitschke, C.; Kiyokawa, K.; Klinker, G.; Takemura, H. Corneal-Imaging Calibration for Optical See-Through Head-Mounted Displays. *IEEE Trans. Vis. Comput. Graph.* **2015**, *21*, 481–490. [CrossRef] [PubMed]
12. Genc, Y.; Tuceryan, M.; Navab, N. Practical solutions for calibration of optical see-through devices. In Proceedings of the IEEE and ACM International Symposium on Mixed and Augmented Reality (ISMAR 2002), Darmstadt, Germany, 30 September–1 October 2002; pp. 169–175. [CrossRef]
13. Makibuchi, N.; Kato, H.; Yoneyama, H. Vision based robust calibration for optical see-through head-mounted displays. In Proceedings of the IEEE International Conference on Image Processing (ICIP 2013), Melbourne, Australia, 15–18 September 2013; pp. 2177–2181. [CrossRef]
14. Owen, C.B.; Zhou, J.; Tang, A.; Xiao, F. Display-relative calibration for optical see-through head-mounted displays. In Proceedings of the IEEE and ACM International Symposium on Mixed and Augmented Reality (ISMAR 2004), Arlington, VA, USA, 2–5 November 2004; pp. 70–78. [CrossRef]
15. Figl, M.; Ede, C.; Hummel, J.; Wanschitz, F.; Ewers, R.; Bergmann, H.; Birkfellner, W. A fully automated calibration method for an optical see-through head-mounted operating microscope with variable zoom and focus. *IEEE Trans. Med Imaging* **2005**, *24*, 1492–1499. [CrossRef] [PubMed]
16. Gilson, S.J.; Fitzgibbon, A.W.; Glennerster, A. Spatial calibration of an optical see-through head-mounted display. *J. Neurosci. Methods* **2008**, *173*, 140–146. [CrossRef] [PubMed]
17. Itoh, Y.; Klinker, G. Performance and sensitivity analysis of INDICA: INteraction-Free DIsplay CAlibration for Optical See-Through Head-Mounted Displays. In Proceedings of the 2014 IEEE International Symposium on Mixed and Augmented Reality (ISMAR 2014), Munich, Germany, 10–12 September 2014.
18. Plopski, A.; Orlosky, J.; Itoh, Y.; Nitschke, C.; Kiyokawa, K.; Klinker, G. Automated Spatial Calibration of HMD Systems with Unconstrained Eye-cameras. In Proceedings of the 2016 IEEE International Symposium on Mixed and Augmented Reality (ISMAR 2016), Merida, Mexico, 19–23 September 2016; pp. 94–99. [CrossRef]
19. Zhang, Z.; Weng, D.; Liu, Y.; Yongtian, W. A Modular Calibration Framework for 3D Interaction System Based on Optical See-Through Head-Mounted Displays in Augmented Reality. In Proceedings of the 2016 International Conference on Virtual Reality and Visualization (ICVRV), Tokyo, Japan, 11–13 May 2016; pp. 393–400. [CrossRef]
20. Klemm, M.; Seebacher, F.; Hoppe, H. High accuracy pixel-wise spatial calibration of optical see-through glasses. *Comput. Graph.* **2017**, *64*, 51–61. [CrossRef]
21. Fontana, U.; Cutolo, F.; Cattari, N.; Ferrari, V. Closed - Loop Calibration for Optical See-Through Near Eye Display with Infinity Focus. In Proceedings of the 2018 IEEE International Symposium on Mixed and Augmented Reality Adjunct (ISMAR-Adjunct 2018), Munich, Germany, 16–20 October 2018; pp. 51–56. [CrossRef]
22. Genc, Y.; Tuceryan, M.; Khamene, A.; Navab, N. Optical see-through calibration with vision based trackers: propagation of projection matrices. In Proceedings of the IEEE and ACM International Symposium on Augmented Reality (ISAR 2001), New York, NY, USA, 29–30 October 2001; pp. 147–156. [CrossRef]
23. Zhang, Z. A flexible new technique for camera calibration. *IEEE Trans. Pattern Anal. Mach. Intell.* **2000**, *22*, 1330–1334. [CrossRef]
24. Hartley, R.; Zisserman, A. *Multiple View Geometry in Computer Vision*; Cambridge books online; Cambridge University Press: Cambridge, UK, 2003.
25. Lee, S.; Hua, H. A Robust Camera-Based Method for Optical Distortion Calibration of Head-Mounted Displays. *J. Disp. Technol.* **2015**, *11*, 845–853. [CrossRef]

26. Holliman, N.S.; Dodgson, N.A.; Favalora, G.E.; Pockett, L. Three-Dimensional Displays: A Review and Applications Analysis. *IEEE Trans. Broadcast.* **2011**, *57*, 362–371. [CrossRef]
27. Trivisio, Lux Prototyping sarl. Available online: https://www.trivisio.com/ (accessed on 24 December 2019).
28. Cutolo, F.; Fontana, U.; Carbone, M.; D'Amato, R.; Ferrari, V. [POSTER] Hybrid Video/Optical See-Through HMD. In Proceedings of the 2017 IEEE International Symposium on Mixed and Augmented Reality (ISMAR-Adjunct 2017), Nantes, France, 9–13 October 2017; pp. 52–57. [CrossRef]
29. Cutolo, F.; Fontana, U.; Ferrari, V. Perspective Preserving Solution for Quasi-Orthoscopic Video See-Through HMDs. *Technologies* **2018**, *6*, 9. [CrossRef]
30. OpenCV, Open Source Computer Vision library. Available online: https://opencv.org/ (accessed on 24 December 2019).
31. Gao, X.-S.; Hou, X.-R.; Tang, J.; Cheng, H.-F. Complete solution classification for the perspective-three-point problem. *IEEE Trans. Pattern Anal. Mach. Intell.* **2003**, *25*, 930–943. [CrossRef]
32. Müller, M.; Rassweiler, M.C.; Klein, J.; Seitel, A.; Gondan, M.; Baumhauer, M.; Teber, D.; Rassweiler, J.J.; Meinzer, H.P.; Maier-Hein, L. Mobile augmented reality for computer-assisted percutaneous nephrolithotomy. *Int. J. Comput. Assist. Radiol. Surg.* **2013**, *8*, 663–675. [CrossRef] [PubMed]
33. Ferrari, V.; Viglialoro, R.M.; Nicoli, P.; Cutolo, F.; Condino, S.; Carbone, M.; Siesto, M.; Ferrari, M. Augmented reality visualization of deformable tubular structures for surgical simulation. *Int. J. Med Robot. Comput. Assist. Surg.* **2016**, *12*, 231–240. [CrossRef] [PubMed]
34. Oishi, T.; Tachi, S. Methods to Calibrate Projection Transformation Parameters for See-through Head-mounted Displays. *Presence Teleoper. Virtual Environ.* **1996**, *5*, 122–135. [CrossRef]
35. Robinett, W.; Rolland, J.P. Computational model for the stereoscopic optics of a head-mounted display. In *Stereoscopic Displays and Applications II*; Merritt, J.O., Fisher, S.S., Eds.; International Society for Optics and Photonics: Bellingham, WA, USA, 1991; Volume 1457; pp. 140–160. [CrossRef]
36. Cattari, N.; Cutolo, F.; D'amato, R.; Fontana, U.; Ferrari, V. Toed-in vs Parallel Displays in Video See-Through Head-Mounted Displays for Close-Up View. *IEEE Access* **2019**, *7*, 159698–159711. [CrossRef]
37. Watson, B.A.; Hodges, L.F. Using texture maps to correct for optical distortion in head-mounted displays. In Proceedings of the Virtual Reality Annual International Symposium '95 (VRAIS '95), Research Triangle Park, NC, USA, 11–15 March 1995; pp. 172–178. [CrossRef]
38. Bax, M.R. Real-time lens distortion correction: 3D video graphics cards are good for more than games. *Stanford ECJ* **2004**, *1*, 9–13. [CrossRef]

© 2019 by the authors. Licensee MDPI, Basel, Switzerland. This article is an open access article distributed under the terms and conditions of the Creative Commons Attribution (CC BY) license (http://creativecommons.org/licenses/by/4.0/).

Article

Pedestrian Detection at Night in Infrared Images Using an Attention-Guided Encoder-Decoder Convolutional Neural Network

Yunfan Chen * and Hyunchul Shin *

Division of Electrical Engineering, Hanyang University, Ansan 426-791, Korea
* Correspondence: chenyunfan@hanyang.ac.kr (Y.C.); shin@hanyang.ac.kr (H.S.);
  Tel.: +82-31-400-4083 (Y.C.); +82-31-400-5176 (H.S.)

Received: 9 December 2019; Accepted: 19 January 2020; Published: 23 January 2020

Abstract: Pedestrian-related accidents are much more likely to occur during nighttime when visible (VI) cameras are much less effective. Unlike VI cameras, infrared (IR) cameras can work in total darkness. However, IR images have several drawbacks, such as low-resolution, noise, and thermal energy characteristics that can differ depending on the weather. To overcome these drawbacks, we propose an IR camera system to identify pedestrians at night that uses a novel attention-guided encoder-decoder convolutional neural network (AED-CNN). In AED-CNN, encoder-decoder modules are introduced to generate multi-scale features, in which new skip connection blocks are incorporated into the decoder to combine the feature maps from the encoder and decoder module. This new architecture increases context information which is helpful for extracting discriminative features from low-resolution and noisy IR images. Furthermore, we propose an attention module to re-weight the multi-scale features generated by the encoder-decoder module. The attention mechanism effectively highlights pedestrians while eliminating background interference, which helps to detect pedestrians under various weather conditions. Empirical experiments on two challenging datasets fully demonstrate that our method shows superior performance. Our approach significantly improves the precision of the state-of-the-art method by 5.1% and 23.78% on the Keimyung University (KMU) and Computer Vision Center (CVC)-09 pedestrian dataset, respectively.

Keywords: infrared pedestrian detection; encoder-decoder; attention; convolutional neural network

---

1. Introduction

Pedestrian detection has attracted considerable attention from researchers in computer vision. Although many studies in pedestrian detection areas have been reported during the past decade [1–8], most of them are confined to detecting pedestrians during daytime using visible (VI) cameras. However, the performance of VI cameras depends on good illumination conditions and can be affected when illumination is poor. Recently, multispectral detectors that employ a fusion of infrared (IR) and VI cameras have been developed to achieve robust and reliable pedestrian detection in adverse illumination circumstances [9–13]. However, the multispectral detectors cannot work well at nighttime since the VI sensor only works when there is a substantial amount of visual information in the environment. Furthermore, the multispectral detectors only support fully aligned images. On a dark night, IR-based pedestrian detectors can effectively replace VI- and multispectral-based detectors, because the IR sensors do not require external light but mainly rely on the radiant temperature of the object. IR pedestrian detection has a wide range of applications, such as patrols, video surveillance, and rescues at night.

The main challenges for robust IR pedestrian detection can be classified into two main types. First, IR images have some adverse properties, such as their noisy nature, low-resolution, and no visual detailed information. These adverse properties make the discriminative feature extraction of an object very difficult, and thus affect the detection performance. Second, IR images are susceptible to weather conditions since IR cameras detect the difference in the temperature of the environment. For instance, pedestrians look brighter than the background in cold weather, while their brightness looks similar to the background in hot weather.

Traditional IR pedestrian detection methods require manually designed features to describe IR objects, which are not conducive to extracting informative features from unstable IR images since IR sensors are susceptible to changeable weather conditions. In recent years, new technologies have been reported with very promising results. Deep learning has enabled progress on object detection using deep convolutional neural networks (CNNs) due to the capacity to generate semantic features via learning from raw pixels. This approach has superior discriminative ability to recognize targets with various shapes from a complex background [14–20]. Therefore, it is very natural to apply the CNN to IR pedestrian detection. A CNN-based method is an effective tool for IR object detection since it can handle variations on images affected by environmental changes, as long as these effects are widely present in the dataset. Therefore, our goal is to design an effective CNN framework specialized for IR pedestrian detection at night time that can achieve the best performance regardless of low-resolution or season changes.

In this research, a new attention-guided encoder-decoder convolutional neural network for accurate IR pedestrian detection at night, called AED-CNN, is developed. The AED-CNN mainly contains an encoder-decoder module and an attention module. The encoder-decoder structure efficiently captures the long-range context information while keeping the spatial information of IR objects, which is helpful for extracting discriminative features from low-resolution and noisy IR images. Based on the encoder-decoder modules, we further propose an attention module to overcome the variation problem due to changeable thermal energy in IR images under different weather conditions.

The following are the main contributions of this work.

- First, novel encoder-decoder modules are proposed to generate multi-scale features. We add an additional decoder module at the end of the single shot multibox detector (SSD) [16] architecture (encoder module) to form an encoder-decoder module, in which a new skip connection block is incorporated into each layer of the decoder to integrate the feature maps from the encoder and decoder modules. The proposed encoder-decoder modules effectively enrich the feature maps via integrating the high-level semantically strong features with low-resolution and low-level detailed features with high-resolution. This method is effective to extract discriminative features even from low-resolution and noisy IR images.
- Second, we propose an attention module that re-weights the multi-scale features generated from the encoder-decoder module. By adding the attention mechanism, the network selectively emphasizes useful information and suppresses ineffective information while re-weighting the features from the encoder-decoder modules. The attention module significantly eliminates background interference while highlighting pedestrians so that there is a boost in the detection performance of the IR pedestrian detector, even when the brightness is similar among the pedestrians and backgrounds.
- Finally, experimental results on two challenging datasets demonstrate that our AED-CNN shows the best performance. Our approach outperforms in detection precision by 5.1% and 23.78% on the Keimyung University (KMU) [21] (the KMU pedestrian detection database [21] is downloaded from: https://cvpr.kmu.ac.kr/ for academic use) and Computer Vision Center (CVC)-09 [22] (the CVC-09 far infrared (FIR) sequence pedestrian dataset [22] (available online, 28 April 2016) is download from: http://adas.cvc.uab.es/elektra/enigma-portfolio/item-1/) pedestrian datasets, respectively, when compared with the state-of-the-art oriented center-symmetric local binary (OCS LBP) + cascade random forest (CaRF) [21] method.

The remaining part of this paper is arranged in the following manners. Section 2 briefly introduces the previous related works. Section 3 explains the proposed AED-CNN in detail. Experimental results and analysis are presented in Section 4. Finally, Section 5 summarizes our work and describes future works.

2. Background

2.1. Infrared (IR) Pedestrian Detection

In the past decades, IR pedestrian detection has attracted much interest from many researchers. Hotspot + SVM [23] acquiesced that a pedestrian's body looks brighter than the background. It generated pedestrian candidates by searching hotspot regions, then a support vector machine (SVM) was applied to conduct the detection stage. On the basis of hotspot features, Ko et al. [24] added analysis of face and shoulder parts in candidate hotspot regions and exploited a random forest classifier to classify the candidates. In [25], a dual-threshold segmentation method was developed to generate regions of interest (ROIs) via detecting hotspot regions. Then the Haar-like and histogram-of-oriented-gradients (HOG) features were extracted for classification. However, the aforementioned hotspot-based methods only show reasonable performance when the thermal energy difference between pedestrians and background is distinguishable. It generates considerable missing instances during hot weather or when the pedestrians wear well-insulated clothing. In [26], a stereo system formed by combining two IR cameras was introduced, in which hotspot detection, edge detection, and disparity calculation were adopted to produce candidate regions. Then, the morphology and thermal features of the pedestrian's head were used to validate candidate regions. In [27], a feature-based region growth with a high-intensity seed method is proposed for segmenting pedestrians, in which vertical deviation-based morphological closing is combined to compensate for distortion caused by clothing. Zhao et al. [28] first proposed a contour precision saliency map for detecting ROI. Then the shape distribution histogram feature was acquired by calculating the distances between the random points on the thinned contour map in the ROI. Finally, a modified sparse representation classification (MSRC) method was developed to detect IR pedestrians. OCS-LBP + CaRF [21] exploited the oriented center-symmetric local binary pattern (OCS-LBP) features to describe the IR pedestrians, and proposed a cascade random forest (CaRF) to classify pedestrians. Biswas et al. [29] proposed a local steering kernel (LSK) to reduce intrinsic noise in IR images and combined an image similarity kernel with SVM to train the LSK tensor. Recently, Heo et al. [30] applied a CNN, named 'you only look once version 2' (YOLOv2) [20] to IR pedestrian detection. They proposed a handcrafted adaptive Boolean-map-based saliency (ABMS) kernel to infuse with YOLOv2. Cao et al. [31] presented an automatic region proposal network by designing a new loss function and adding a segmentation task, for IR pedestrian detection. This method does not consider the factors that IR images are susceptible to environmental changes, and thus the results are not ideal. In the latest work [32], a CNN-based IR person detector based on residual network (ResNet) [33] and atrous spatial pyramid pooling (SPP-net) [34] was proposed. The developed IR person detector was evaluated by using infrared closed-circuit television (CCTV) images.

2.2. Convolutional Neural Network (CNN)-Based Object Detection

With the rapid growth of CNN technologies recently, a large variety of CNN-based approaches have facilitated object detection to a new stage. CNN-based object detectors are generally grouped into two main categoreis. The first categore is termed as two-stage approaches, including fast region-based CNN (Fast R-CNN) [14] and Faster R-CNN [15]. The two-stage methods include two parts. The first part aims to generate a sparse series of candidate object proposals, and the second part refines the candidate proposals to determine the accurate objects locations and its corresponding class label. Although the aforementioned two-stage detectors achieve desirable accuracy, the computational speed is slow. By contrast with two-stage detectors, one-stage methods such as SSD [16] and YOLO [20,35] avoid generating region proposals and resampling features by encapsulating all operations in a single network, which performs better in speed than the two-stage detectors. In CNN-based methods, the high-level layers which contain sufficient semantic features lack the detailed spatial features of the objects because of the striding operations in pooling and convolutional layers. This makes it hard to extract discriminative features. Consequently, it is difficult to identify target locations. On the contrary, the features from the low-level layers which contain enough spatial information but lack the semantic information are not robust enough to the challenges of appearance variation and occlusions. To address these issues, some networks [18,36] try to observe and utilize the pyramidal features to a large extent by building an encoder-decoder architecture with lateral connections. These networks show dramatic improvements in accuracy when compared with conventional detectors. However, effective methods to combine the information from the encoder and decoder modules for good performance still need to be explored. Motivated by the aforementioned works, we introduce a novel encoder-decoder module that takes an IR image as an input to the encoder module to capture richer semantic information and adopts a decoder module to progressively recover the spatial information. In our proposed encoder-decoder module, a new skip-connection block is used to fuse the information from the encoder and decoder module. The encoder-decoder structure can efficiently capture the long-range context information while keeping the spatial information of IR objects. This owes to the expressiveness of the encoder-decoder layers, which are originated from the combinatorial nature of Rectified Linear Unit (ReLU) for decomposition and reconstruction.

2.3. Attention Mechanism

Attention mechanism is widely utilized in CNNs to solve different computer vision tasks, like video classification [37] and object detection [38]. Wang et al. [37] applied the attention mechanism to non-local filter operation, which is effective in performing classification tasks in videos through computing the feature relationships in different positions. DeepSaliency [38] exploited the attention mechanism to learn the spatial relationships between salient features, which shows promising results for object detection and segmentation tasks. The above works are to investigate to model spatial correlations. In contrast, another method, squeeze-and-excitation network (SENet) [39], is to model the correlative dependence between channels of CNN features to perform feature recalibration. It utilizes global information to learn layout or emphasis of the scene and to obtain the context information which guides visual processing to task-related image regions. Inspired by SENet [39], we find that we can recalibrate the features in the decoder aiming at guiding the detector to pay more attention to pedestrian locations. Specifically, we propose an attention module that adopts SE blocks to learn attention vectors from the feature maps in the encoder. Then the learned attention vectors are utilized to scale the feature maps from the decoder to enhance informative features and to suppress less useful features. The attention module indicates the probability that each pixel belongs to a pedestrian region. It has a high discriminative capability to recognize a human target, because background interference is reduced, and pedestrians are highlighted.

In this research, a novel attention-guided encoder-decoder convolutional neural network (AED-CNN) is developed to detect pedestrians in IR images. To better describe pedestrians in low-resolution and noisy IR images, the encoder-decoder module is adopted, in which a skip connection block is integrated to effectively combine the features from the encoder and decoder for increasing additional context information. Furthermore, the attention module is proposed to highlight pedestrians while eliminating background interference, which can help to detect pedestrians under different weather conditions.

3. Proposed Attention-Guided Encoder-Decoder Convolutional Neural Network (AED-CNN)

The proposed AED-CNN mainly contains an encoder-decoder module and an attention module. The encoder-decoder module takes an IR image as an input to the encoder module to capture rich semantic information and adopts a decoder module to progressively recover the spatial information. The attention module learns attention vectors from the feature maps in the encoder, which are used to scale the feature maps from the decoder. The attention module effectively enhances informative

3.1. Overview of the Proposed AED-CNN Architecture

Figure 1 presents an overview of our AED-CNN architecture. The input image is processed by the encoder module. The encoder gradually reduces the resolution of the feature map for conducting a multi-scale search of bounding boxes, as in the SSD [16]. Then, the feature maps in the decoder are up-scaled via deconvolutional layers and then combined with the corresponding feature maps of the same resolution in the encoder through skip connection blocks. The multi-scale features generated by the encoder-decoder module are re-weighted through the attention net. Finally, these re-weighted features are sent to the prediction stage for classifying pedestrians and regressing bounding boxes. The architectures of skip connection block and the attention net are presented in Figures 2 and 3, respectively.

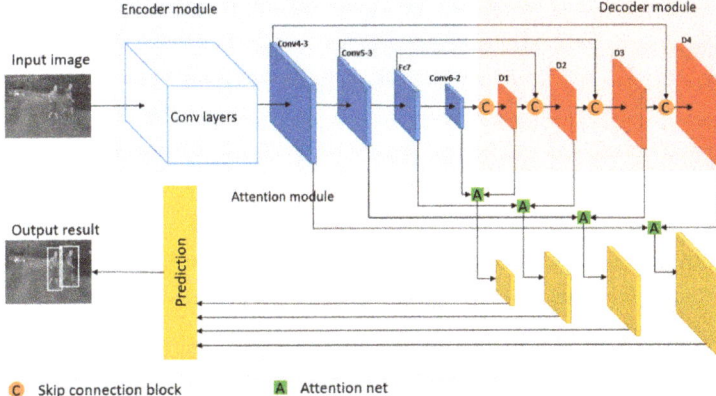

Figure 1. Architecture of the proposed attention-guided encoder-decoder convolutional neural network (AED-CNN).

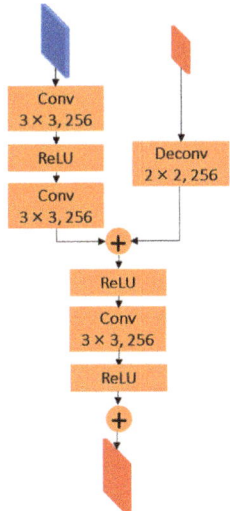

Figure 2. Skip connection block.

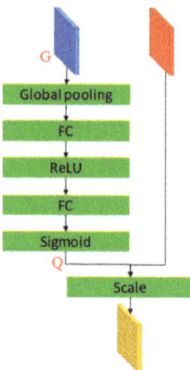

Figure 3. Attention net.

3.2. Encoder Module

As shown in Figure 1, using a similar method as in SSD [16], the encoder module is built on top of a "base" network (a feedforward convolutional network that contains several convolution layers and pooling layers) which is followed by a series of progressively smaller convolution layers to constitute a cascade of feature maps with a gradually increasing field of view and decreasing spatial resolution. The VGG-16 [40] is chosen as our "base" network since it has good detection performance as well as fast detection speed. Following [19], two additional convolution layers (conv6-1 and conv6-2) are added after fully connected layer (Fc7) of the truncated VGG-16 to obtain more semantic information at a high-level and to detect pedestrians at multi-scales. However, it is hard for SSD to classify the pedestrians in IR images of low-resolution and with noisy characteristics, owing to the weak semantic information on the shallow features. Inspired by [18], we added a decoder module in the proposed architecture to improve the quality of IR images. The decoder module can increase more context information that enriches semantic features extracted from IR images. The decoder module is described in the next section.

3.3. Decoder Module

In the encoder module, the low-level layers contain high spatial resolution but lack enough semantic features. By contrast, the high-level layers have rich semantic features but low spatial resolution. It is difficult to extract discriminative features from the feature layers of encoder module to detect pedestrians in low-resolution IR images. In [18], the deconvolutional network has been proposed to refine the traditional feedforward convolutional network, which shows that the deconvolutional module equalizes the representability of each feature map of a sophisticated network and makes the network more efficient. Therefore, to include more high-level context in detection, we add a decoder module at the end of the encoder to effectively make an encoder-decoder hourglass network structure. The decoder module consists of a series of deconvolution layers with successively increasing resolution of feature map layers, namely deconvolution layer 1 (D1) to deconvolution layer 4 (D4), as shown by the red blocks in Figure 1. In [18], five deconvolution layers are used from the top-most feature maps, in which all fine details are missing. Moreover, the additional deconvolution layers result in more computational cost, making it impractical for real-time applications because of the large inference time. To overcome this, we make our decoder shallower and only four layers are used to maintain a higher detection speed.

Skip Connection Block

As mentioned above, the encoder-decoder module is exploited to obtain feature maps with rich context information via combining the high-level semantic information with the low-level detailed information. To this end, a new skip connection block, as shown in Figure 2, is introduced to fuse the feature maps from the decoder and the corresponding feature maps from the encoder. In order to share the structure of skip connection block, the deep layers are symmetrically connected with these shallow layers. In other words, the deep features have the same down sampling factor. Specifically, feature maps in the decoder are up sampled and then merged with the feature maps in the decoder. Figure 2 shows the proposed skip connection block. First, the feature map in the encoder is up sampled by a 2 × 2 deconvolution layer to match the size of the feature map in the decoder. The feature map in the decoder is followed by a 3 × 3 convolution layer, a ReLU layer, and a 3 × 3 convolution layer. We merge them by using the element-wise sum. Then, the summed layer passes through a block (ReLU, 3 × 3 convolution, ReLU), which is useful for discernable feature extraction as proven in [19], and then summed with the deconvolution layer, which skips the block. Note that the Conv6-2 directly goes through the first skip connected block to generate D1 without combination with the feature map in the encoder module. The skip connection is beneficial to back-propagation, which can effectively solve the problem of performance degradation of deep convolutional neural networks under extremely deep conditions, as shown in [33]. It speeds up the training of a deeper network and thus makes the learning easier. In addition, the skip connection helps traverse the information in deep neural networks to boost the classification performance.

3.4. Attention Module

The attention module plays an important role in AED-CNN to re-weight the multi-scale features. As shown in the lower part of Figure 1, the attention module contains four attention nets which take the feature maps of the same scale in the encoder and decoder as inputs, and then outputs four re-weighted feature maps in multi-scales for the final pedestrian prediction part. Under different weather conditions, the difference in thermal energy between pedestrian and background will change. In hot weather, pedestrians' brightness looks similar to the background, which makes it difficult for pedestrians to be distinguished. The success of SENet [39] shows the effectiveness of the attention mechanism on images classification. SENet proposed a channel-wise attention mechanism that helps to selectively enhance useful features and suppress ineffective features by performing feature recalibration. For IR pedestrian detection, we need to find the most effective features to highlight IR pedestrian regions. Therefore, we propose attention net to recalibrate the multi-scale features generated by the decoder module, in which large weights are assigned to channels that show a high response to IR objects. The key idea of the attention net is to learn the feature weights according to the loss, with the objective of maximizing the weights of effective feature maps and minimizing the weights of less effective feature maps. The attention mechanism filters out some background details according to re-weighted features and focuses more on the foreground regions, which helps to generate effective features for pedestrian detection.

Figure 3 shows the architecture of our attention net. The left side of Figure 3 takes the feature map in the encoder as the input guidance G. Then the guidance G passes through one global average pooling layer and two fully connected (FC) layers (The two FC layers are followed by a ReLU layer and a sigmoid layer, respectively) to learn the mapping function F used to generate the channel-wise attention vector Q:

$$Q = F(G^T) \qquad (1)$$

The generated attention vector Q is used to scale the layer of the decoder. We apply the same attention net to four groups of feature maps of the same scale from the encoder and decoder. Finally, four multi-scale re-weighted layers are generated, which will be used for the final pedestrian prediction task. Figure 4 shows a visualization of feature maps with and without the attention module. One can see that the pedestrian regions are highlighted while the backgrounds are suppressed, in the feature maps from the attention module.

The attention module specifically models the interdependencies between the convolutional channels that effectively boost the representational capability for various samples. After applying the attention module, more useful features are emphasized while less informative ones are restrained, via re-weighting the sample-reliant features. In other words, the pedestrians are highlighted, while the background interferences are suppressed. This ensures the performance of the detector when the thermal energy of pedestrians is not distinguishable due to weather changes. The attention module is not only easy to implement but also realizes remarkable improvements at little extra computational cost.

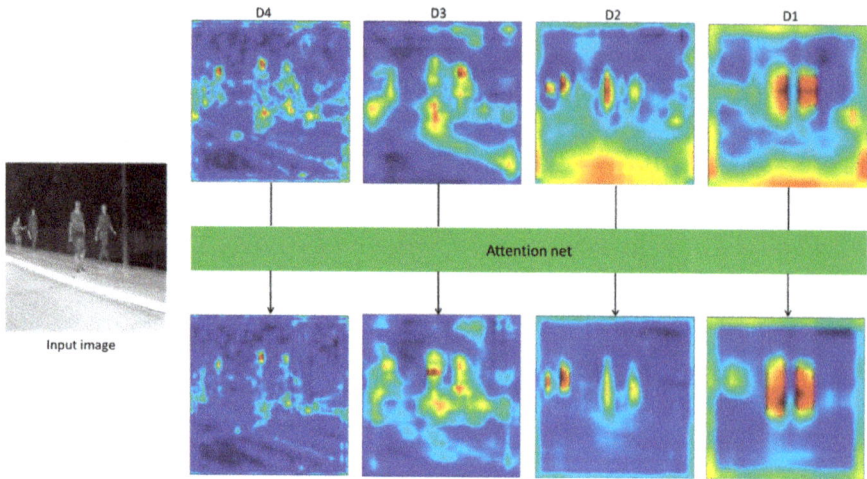

Figure 4. Visualization of feature maps from decoder module (**top**) and visualization of feature maps from the attention module (**bottom**).

3.5. Training

3.5.1. Matching and Hard Negative Mining

In the training stage, the anchor boxes are used to localize objects in multiple scales, which need to be matched with the ground truth bounding boxes. The correspondence between each ground truth bounding box A to the anchor box B is determined by the jaccard overlap [41]. The value of the jaccard overlap is defined as:

$$J(A, B) = \frac{\text{area}(A \cap B)}{\text{area}(A \cup B)} \quad (2)$$

Each ground truth bounding box is first matched to the anchor box with the best $J(A, B)$. Then the remaining anchor boxes are matched to any ground truth with $J(A, B) > 0.5$. This strategy is beneficial to predict multiple bounding boxes with high scores for overlapped objects. After matching, the majority of samples are negative. Similar to SSD [16], we select the negative samples with the top loss values from the non-matched anchor boxes to set the ratio between positive and negative samples as 1:3.

3.5.2. Loss Function

Equation (3) shows our overall loss function, which is a weighted summation of the two branches, one is the confidence loss (*conf*) of the softmax classifier, and the other is the localization loss (*loc*) of the bounding box regression.

$$L(\{p_i\}, \{t_i\}) = \frac{1}{N}\left(\left(L_{conf}\left(p_i, p_i^*\right)\right) + \lambda L_{loc}\left(t_i, t_i^*\right)\right) \quad (3)$$

where p_i^* is the ground truth label of an anchor i in a mini-batch, and the value of p_i is the probability of an anchor i being a pedestrian. The ground truth location of the anchor i is denoted by t_i^* and the predicted bounding box location of the anchor i is represented by t_i^*. N denotes the number of positive anchors. Notably, if $N = 0$, the overall loss $L = 0$. Currently, we set weight term λ to 1 via cross-validation. The classification loss L_{conf} is the cross-entropy loss over two classes (pedestrian vs. non-pedestrian). As in Fast R-CNN [14], the smooth L_1 loss is adopted as our regression loss L_{loc}.

$$L_{loc}(t_i, t_i^*) = \sum_{j \in \{x,y,w,h\}} smooth_{L_1}(t_j, t_j^*) \tag{4}$$

where $smooth_{L_1}$ is defined as:

$$smooth_{L_1}(x) = \begin{cases} 0.5x^2 & if\ |x| < 1 \\ |x| - 0.5 & otherwise, \end{cases} \tag{5}$$

3.5.3. Optimization

The "base" network VGG-16 in our AED-CNN is pretrained on the ImageNet Large Scale Visual Recognition Challenge (ILSVRC) dataset [42]. The Xavier approach [43] is applied to initialize the parameters of two additionally convolutional layers (conv6-1 and conv6-2). In training, the default batch size is set to 11. Then, the entire network is fine-tuned by using a stochastic gradient descent method with the momentum 0.9 and weight decay 0.0005. To prevent gradient explosion in early iterations, we first run 10 k (where k = 1000) iterations using a learning rate of 0.00005. Then, we reset the learning rate to 0.001 for the next 70 k iterations. After the completion of 70 k iterations, the learning rate is reduced by 10 times after every 20 k iterations. Learning stops after 120 k iterations.

4. Experimental Results

4.1. Datasets and Processing Platform

Even though there are extensive color video/image datasets for pedestrian detection, only a small number of thermal infrared pedestrian datasets are available. In this study, we evaluate our proposed AED-CNN on the widely used KMU [21] and CVC-09 [22] pedestrian datasets.

The KMU was captured by a far infrared (FIR) camera from moving vehicles (at 20 to 30 km/h) during the summer and winter nights for pedestrian detection. The training data contains 4474 positive images and 3405 negative images. The positive images include pedestrians with various sizes and postures. The negative images were generated through random cropping from the background. The testing data contains 5045 images of the same size of 640 × 480, including pedestrians with various activities under different weather conditions, such as walking down the sidewalk or crossing the road in hot summer as well as in cold winter.

The CVC-09 FIR sequence pedestrian dataset was collected using a FIR camera mounted on a car roof during summer days. This dataset is more challenging since it comprises pedestrians with varying moving speeds, a variety of motions, all types of poses, changeable thermal energy, and partial or full occlusions at night. The CVC-09 contains 5309 positive images and 2115 negative images for training, 5763 images for testing, with the same size of 640 × 480.

For both KMU and CVC-09 datasets, the pedestrian images of different scales are combined for training. The experiments are performed on a standard computer under Ubuntu 14.04 with a core i7-6850k 3.6 GHz central processing unit (CPU) and 64 GB random-access memory. For the graphics processing unit (GPU), we used three NVIDIA Titan X for training and a single Nvidia Titan X for testing. Our codes are built on Caffe which requires a compute unified device architecture (CUDA) and CUDA deep neural network library (cuDNN).

4.2. Evaluation Metric

We validate the detection performance by employing precision-recall curves, which are generally applied to the evaluation of human detection performance. To verify the detection results, the predicted bounding boxes generated by the detector are compared with the ground-truth bounding boxes and checked as either true positive (TP), false positive (FP), or false negative (FN), via measuring the overlap between the bounding boxes. TP indicates the quantity of properly detected pedestrians, FP indicates the number of mistakenly detected pedestrians by the detector, and FN indicates the number of pedestrians that are not detected by the detector. As the detection criteria, the detected bounding box is considered to be TP if the overlap ratio between the detected bounding box and the ground-truth bounding box exceeds 50%. TP/(TP + FP) calculates the precision and TP/(TP + FN) computes the recall. The average precision (AP) describes the tendency of the precision-recall curve and is calculated by averaging the precision at several evenly spaced recall levels via changing the threshold of the detection scores. In our experiments, the AP is obtained by averaging the precision values at 11 evenly spaced recall levels between 0 and 1.

4.3. Comparison of the Detection Performance on the Keimyung University (KMU) Dataset

We compare the performance of AED-CNN with a set of five well-known methods, including histogram-of-oriented-gradients (HOG) + support vector machine (SVM) [44], Hotspot + SVM [23], OCS-LBP + CaRF [21], YOLOv2 [20], and YOLOv2 + ABMS [30].

As shown in Figure 5, our AED-CNN clearly shows the best performance when compared to other methods and achieves the highest AP of 97.5%, which significantly exceeds the two recent optimal results, YOLOv2 + ABMS [30] by 17%, and OCS-LBP + CaRF [21] by 5.1%, respectively. HOG + SVM [44] shows the worst results because the HOG feature is not suitable to characterize pedestrians in IR images. Although the hotspot regions of pedestrians can be detected in IR images, Hotspot + SVM [23] is still ineffective when the temperature of pedestrians is similar to the background, for example during a summer night. With the same limitations as Hotspot + SVM, OCS-LBP + RF [21] and YOLOv2 [20] showed worse performance than ours. ABMS is proposed by YOLOv2 + ABMS [30] to pre-process the IR images for enhancing pedestrians during hot weather. However, the YOLOv2 network cannot detect small-sized pedestrians in low-resolution IR images. These comparative results prove that our AED-CNN shows superior robustness compared to other methods under various circumstances.

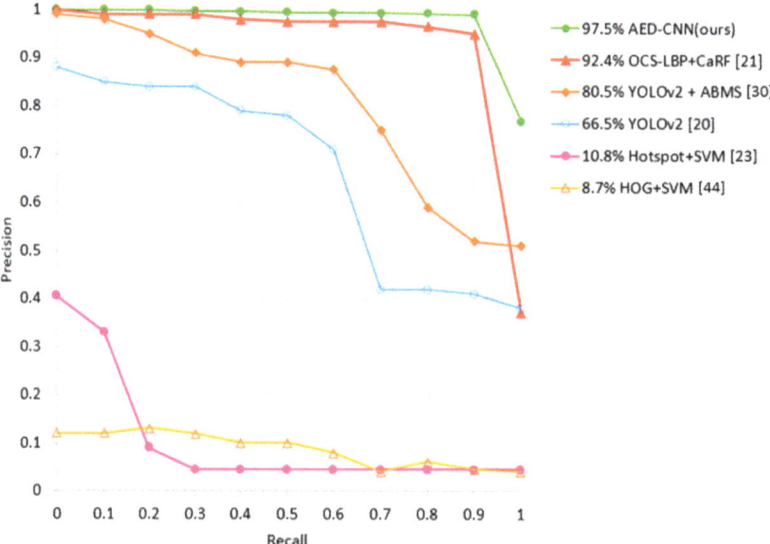

Figure 5. Performance comparison of precision versus recall using the Keimyung University (KMU) dataset.

Figure 6 presents some detection results on the KMU test set using our method. It is clear that our method successfully detects pedestrians with various scale instances during a hot summer night. From this observation, we can conclude that our method is effective at detecting pedestrians in IR images and is robust under different weather conditions.

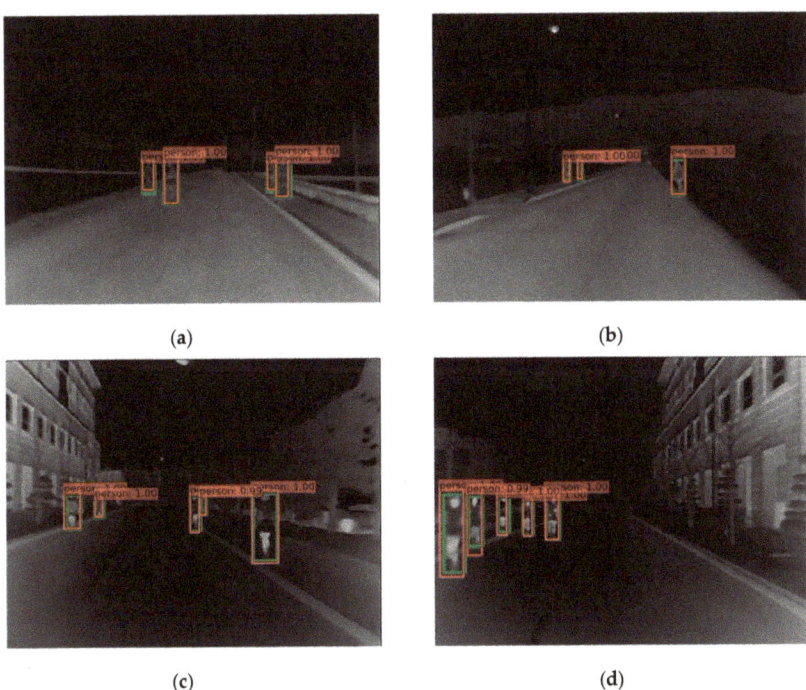

Figure 6. Some detection results on the KMU test dataset. The green bounding boxes denote the ground truth, and the red bounding boxes show the detection results. (**a**–**d**) are visualization of detection results on the image set selected from the dataset.

4.4. Comparison of the Detection Performance on the CVC-09 Dataset

The detection performance is also evaluated by using the CVC-09 dataset. The precision-recall curves of the proposed AED-CNN and other well-known approaches are shown in Figure 7. A similar tendency can be observed in the results using the KMU dataset: the AP gap is quite large, 87.68% of ours versus 63.9% of the state-of-the-art OCS-LBP + CaRF [21]. Figure 7 reveals that our approach significantly outperforms all other approaches. The results are meaningful because the CVC09 dataset includes numerous pedestrian images with thermal energy levels similar to the background levels. Furthermore, the resolution is low, and some images are occluded. The performance comparison in Figure 7 demonstrates that our AED-CNN is obviously better than other state-of-the-art methods in detecting pedestrians.

Example detection results of our method on the CVC09 test set are displayed in Figure 8. In Figure 8a,b, AED-CNN can detect all the pedestrians. Figure 8c shows that a heavily occluded pedestrian was missed. In Figure 8d, three heavily occluded pedestrians were missed. From these observations, we think that it is necessary to improve the performance of AED-CNN to detect pedestrians with heavy occlusions in the future, even though AED-CNN shows the best performance when compared with other state-of-the-art methods.

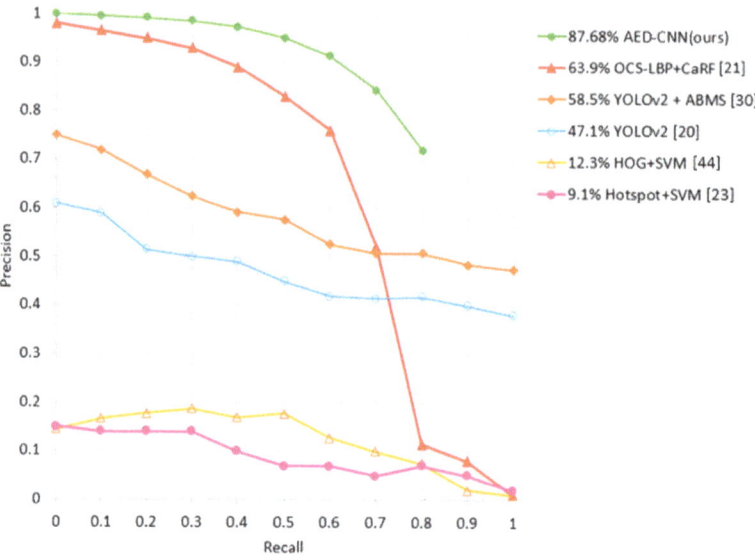

Figure 7. Performance comparison of precision versus recall by using the Computer Vision Center (CVC)09 dataset.

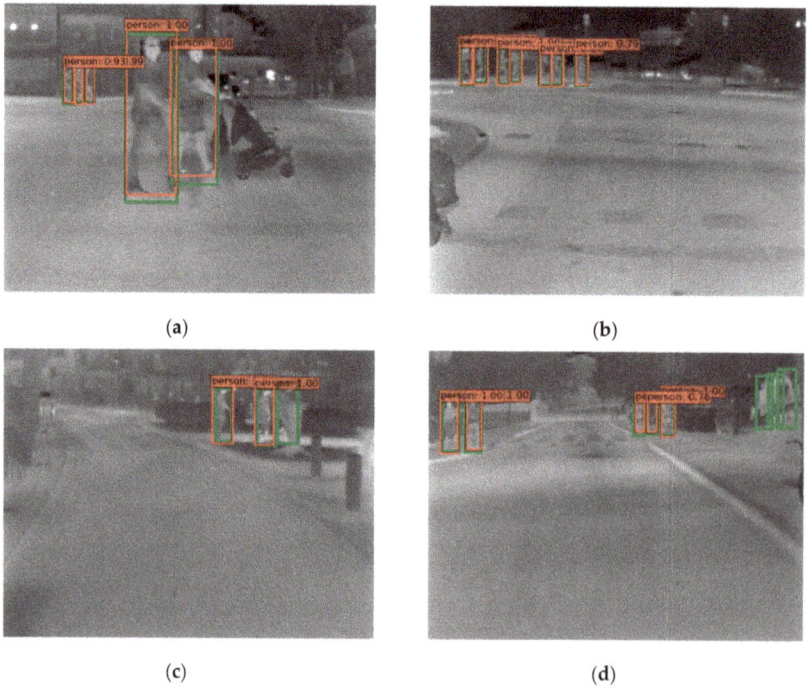

Figure 8. Four detection results of the CVC09 test dataset. The green bounding boxes denote the ground truth, and the red bounding boxes show the detection results. (**a**–**d**) are visualization of detection results on the image set selected from the dataset.

4.5. Comparison of the Computational Speed

The computational speed and AP of the proposed AED-CNN with other reported methods are compared in Table 1. We measure the running times of all the methods using the same machine. Although YOLOv2 [20] and YOLOv2 + ABMS [30] have the fastest computational speed (0.02 s/f), their precision is worse than our approach (over 10%). The computational speed of our method is 0.03s per image, which is very competitive compared to other approaches.

Table 1. Comparison of the computation times and average precisions (APs) for the Keimyung University (KMU) and Computer Vision Center (CVC)09 test sets.

Methods	KMU Test Set		CVC09 Test Set	
	Speed (s/f)	Precision	Speed (s/f)	Precision
Histogram-of-oriented-gradients (HOG) + Support vector machine (SVM) [44]	0.09	8.7%	0.09	12.3%
Hotspot + SVM [23]	0.08	10.8%	0.08	9.1%
Oriented center-symmetric local binary (OCS-LBP) + Cascade random forest (CaRF) [21]	0.06	92.4%	0.06	63.9%
You only look once version 2 (YOLOv2) [20]	0.02	66.5%	0.02	47.1%
YOLOv2 + Adaptive Boolean-map-based saliency (ABMS) [30]	0.02	80.5%	0.02	58.5%
Attention-guided encoder-decoder convolutional neural network (AED-CNN) (ours)	0.03	97.5%	0.03	87.68%

4.6. Ablation Experiments

The ablation experiments are typically used to remove the main proposed components of the developed method gradually to evaluate the effectiveness of each component on detection performance. To verify the effectiveness of the encoder-decoder module and attention module, ablation experiments have been done using the KMU dataset. The initial simulation is the original SSD [16], which is the baseline detector. Then the decoder module is added to make an improved version, which is the encoder-decoder module. Finally, the attention module is added to form our AED-CNN. The average precision value of each module is given in Table 2. The baseline detector SSD [16] achieved 92.04% AP, while the encoder-decoder module and attention module have significantly improved the detection accuracies, of 95.36% and 97.50%, respectively.

Table 2. Results of the ablation experiments on KMU test set.

Methods	Average Precision on KMU Test Set
Single shot multibox detector (SSD) [16]	92.04%
Encoder-decoder module	95.36%
Encoder-decoder module + attention module (AED-CNN)	97.50%

4.6.1. Evaluation of the Encoder-Decoder Module

As shown in Table 2, the encoder-decoder module significantly outperforms the original SSD [16], by improving the AP by 3.32%. The SSD [16] built the pyramid starting from high up in the network without using low-level feature maps. However, the low-level feature maps contain high spatial resolutions and rich detailed features, which is essential for pedestrian detection in low-resolution IR images. Our method proposed a new decoder module to effectively combine the high-resolution low-level detailed features with the low-resolution high-level semantic features, which can increase context information that helps to boost the pedestrian detection rate in low-resolution IR images. Figure 9a,b illustrate visual comparisons of the original SSD detection results versus the encoder-decoder detection results. It is clear that the SSD produces more false positives and false negatives.

(a) (b) (c)

Figure 9. Visual comparisons of detection results on four example images among SSD, encoder-decoder module, and encoder-decoder + attention (AED-CNN). The green bounding boxes denote the ground truth. The red bounding boxes show the detection results. (**a**) SSD [16], (**b**) encoder-decoder module (**c**) encoder-decoder + attention (AED-CNN).

4.6.2. Evaluation of the Attention Module

By incorporating our proposed attention module with the encoder-decoder module, the detection AP is further improved by 2.14%, from 95.36% to 97.05%, as shown in Table 2. This comparison demonstrates that the proposed attention module is useful to detect pedestrians in IR images with changeable thermal energy by highlighting pedestrian regions while reducing background interference. The visual comparison of the encoder-decoder detection results and those of the AED-CNN detection is provided in Figure 9b,c. One can see that the encoder-decoder module without the attention module fails to detect pedestrians, when the brightness values of pedestrians are close to those of the background. However, the AED-CNN is able to successfully detect all pedestrians in these examples even when the thermal energy difference between pedestrians and background is indistinguishable.

4.7. Discussion

The objective of this research is to solve two main challenges of IR pedestrian detection. One is that IR images have some adverse properties, such as a noisy nature and low-resolution. The other is that IR images are susceptible to weather. To overcome these two problems, we propose an attention-guided encoder-decoder convolutional neural network. Motivated by the recent success of CNNs in object detection, we design a novel encoder-decoder module based on the well-known SSD approach [16]. The proposed encoder-decoder module learns effective features from low-resolution IR images. Inspired by the attention mechanism applied in CNN [39], we further propose an attention module to highlight IR pedestrians even when thermal energy is indistinguishable between pedestrians and the background.

4.7.1. Explanation of Results

Empirical experiments described in Section 4.6 well validate the effectiveness of the proposed encoder-decoder and attention modules. Each of these modules in AED-CNN can bring a significant improvement in the detection accuracy. In Sections 4.3 and 4.4, we compare our method with several published state-of-the-art methods, on two challenging datasets. From the experimental results, we found our proposed CNN-based architecture significantly outperforms hand-crafted methods. The proposed method outperforms the hand-crafted method by 5.1% and 23.78% in AP on the KMU and CVC-09 pedestrian dataset, respectively. This result inspires us to explore in depth how to develop more effective CNN architectures to further improve the performance of IR pedestrian detection. As far as I know, there are only few research works which apply CNN to IR pedestrian detection.

Another research direction we can further explore is how to improve the confidence score of detected pedestrians in adverse environments. One can compare the result visualization of Figures 6 and 8, to easily find that the confidence score in Figure 8 is lower. The confidence score of some detected pedestrians in Figure 8b,d are about 0.7. The reason is that the difference in brightness between pedestrians and the background in the CVC-09 data is lower than the KMU data. It is necessary to explore how to further emphasize pedestrian regions. Some recent works apply CNN for saliency detection in images [45], which shows promising results. Saliency detection aims to learn the significant difference between different groups. Saliency detection helps to reduce the complicity of the background and detect the foreground objects. For the task of IR pedestrian detection, the pedestrian and background are two groups, we can apply saliency detection algorithms to learn specific features to find pixels that belong to saliency regions (pedestrian regions) [45]. Therefore, how to incorporate saliency detection into IR pedestrian detection can be a future research problem.

4.7.2. Limitation of the Proposed Method

Despite having achieved state-of-the-art detection accuracy and real-time detection speed, our approach does not perform well in some cases, as shown in Figure 10. The failure cases are caused by occlusions. In Figure 10a, one person occluded by a tree is not detected. In Figure 10b, one person in crowed is not detected. In Figure 10c, a false detection is generated among the crowd. Thus, occlusions may degrade detection performance, which is a common limitation with any methods.

It is necessary to solve the challenges of detecting occluded pedestrians. This common problem has already been explored [46,47]. We can regard pedestrians as a combination of different body parts. For each body part, CNN learns features and generate a score, respectively. If the body part is occluded, the score will be lower, otherwise, the score will be higher. During training, the features of each body part will be integrated. The integrated feature will be used for pedestrian classification and localization. The issues to solve in future are how many parts of a body should be used and how to effectively integrate the features of each body part, with low resolution IR images. In addition, we can take advantage of supervised learning to optimize the occlusion problem from the loss function perspective. We can consider designing the loss function to force proposals away from the second largest ground-truth bounding boxes that overlap with it. The objective is to make the proposal close to real objects and away from the false objects, thereby reducing the false detection rate due to occlusions.

 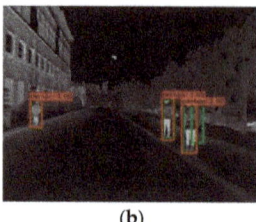 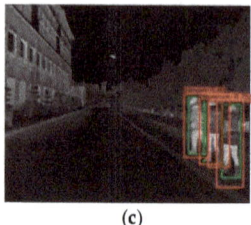

(a)   (b)   (c)

Figure 10. Failure cases of our proposed infrared (IR) pedestrian detection method. The green bounding boxes denote the ground truth. The red bounding boxes show the detection results. (**a**–**c**) are visualization of failure cases on the image set selected from the KMU and CVC09 datasets.

5. Conclusions

In this paper, an effective attention-guided encoder-decoder convolutional neural network (AED-CNN) is developed for pedestrian detection at night using IR images with various road scenes and weather conditions. The encoder-decoder module is used for generating multi-scale feature maps, in which the skip-connection block is proposed to combine the features from the encoder and decoder layers. This combination increases context information that is helpful to extract discriminative features even in low-resolution and noisy IR images. Furthermore, an attention module is devised to highlight pedestrians while eliminating background interference, which is effective to detect pedestrians even when the differences in thermal energy between pedestrians and background is poorly distinguishable.

Experimental results on the two widely used datasets fully verify that our method significantly outperforms other state-of-the-art approaches. The proposed method achieves 97.50% AP on the well-known KMU pedestrian dataset, which is 5.1% better in detection accuracy and two times faster in detection speed than the published state-of-the-art method. Furthermore, our method achieves an AP of 87.68% which is a 23.78% improvement over the previous best performance on the challenging CVC-09 pedestrian dataset. Moreover, our AED-CNN achieves a real-time detection speed of about 0.03 s/f. In future works, solutions to detect occluded IR pedestrians will be explored.

Author Contributions: Y.C. developed the idea, implemented the experiments, and wrote the manuscript. H.S. supervised the research and performed revisions and improvements. All authors have read and agreed to the published version of the manuscript.

Funding: This research was funded by National Research Foundation of Korea (2017-R1D1A1B04-031040).

Acknowledgments: This work was supported by Basic Research Project in Science and Engineering through the Ministry of Education of the Republic of Korea and National Research Foundation of Korea (National Research Foundation of Korea 2017-R1D1A1B04-031040).

Conflicts of Interest: The authors declare no conflicts of interest.

References

1. Dollár, P.; Appel, R.; Belongie, S.; Perona, P. Fast feature pyramids for object detection. *IEEE Trans. Pattern Anal.* **2014**, *36*, 1532–1545. [CrossRef]
2. Tian, Y.; Luo, P.; Wang, X.; Tang, X. Deep learning strong parts for pedestrian detection. In Proceedings of the IEEE International Conference on Computer Vision, Santiago, Chile, 7–13 December 2015; pp. 1904–1912.
3. Cai, Z.; Fan, Q.; Feris, R.S.; Vasconcelos, N. A unified multi-scale deep convolutional neural network for fast object detection. In Proceedings of the European Conference on Computer Vision, Amsterdam, The Netherlands, 11–14 October 2016; pp. 354–370.
4. Xiao, F.; Liu, B.; Li, R. Pedestrian object detection with fusion of visual attention mechanism and semantic computation. *Multimed. Tools Appl.* **2019**, 1–15. [CrossRef]
5. Brazil, G.; Yin, X.; Liu, X. Illuminating pedestrians via simultaneous detection & segmentation. In Proceedings of the IEEE International Conference on Computer Vision, Venice, Italy, 22–29 October 2017; pp. 4950–4959.

6. Guo, Z.; Liao, W.; Xiao, Y.; Veelaert, P.; Philips, W. An occlusion-robust feature selection framework in pedestrian detection. *Sensors* **2018**, *18*, 2272. [CrossRef]
7. Li, J.; Liang, X.; Shen, S.; Xu, T.; Feng, J.; Yan, S. Scale-aware fast r-cnn for pedestrian detection. *IEEE Trans. Multimed.* **2017**, *20*, 985–996. [CrossRef]
8. Gu, J.; Lan, C.; Chen, W.; Han, H. Joint pedestrian and body part detection via semantic relationship learning. *Appl. Sci.* **2019**, *9*, 752. [CrossRef]
9. Hwang, S.; Park, J.; Kim, N.; Choi, Y.; So Kweon, I. Multispectral pedestrian detection: Benchmark dataset and baseline. In Proceedings of the IEEE Conference on Computer Vision and Pattern Recognition, Boston, MA, USA, 7–12 June 2015; pp. 1037–1045.
10. Liu, J.; Zhang, S.; Wang, S.; Metaxas, D.N. Multispectral deep neural networks for pedestrian detection. In Proceedings of the British Machine Vision Conference, York, UK, 19–22 September 2016; pp. 1–13.
11. König, D.; Adam, M.; Jarvers, C.; Layher, G.; Neumann, H.; Teutsch, M. Fully convolutional region proposal networks for multispectral person detection. In Proceedings of the IEEE Conference on Computer Vision and Pattern Recognition Workshops, Honolulu, HI, USA, 21–26 July 2017; pp. 243–250.
12. Chen, Y.; Xie, H.; Shin, H. Multi-layer fusion techniques using a CNN for multispectral pedestrian detection. *IET Comput. Vis.* **2018**, *12*, 1179–1187. [CrossRef]
13. Li, C.; Song, D.; Tong, R.; Tang, M. Illumination-aware faster r-cnn for robust multispectral pedestrian detection. *Pattern Recognit.* **2019**, *85*, 161–171. [CrossRef]
14. Girshick, R. Fast r-cnn. In Proceedings of the IEEE International Conference on Computer Vision, Santiago, Chile, 7–13 December 2015; pp. 1440–1448.
15. Ren, S.; He, K.; Girshick, R.; Sun, J. Faster r-cnn: Towards real-time object detection with region proposal networks. In Proceedings of the Advances in Neural Information Processing Systems, Montreal, QC, Canada, 7–12 December 2015; pp. 91–99.
16. Liu, W.; Anguelov, D.; Erhan, D.; Szegedy, C.; Reed, S.; Fu, C.Y.; Berg, A.C. Ssd: Single shot multibox detector. In Proceedings of the European Conference on Computer Vision, Amsterdam, The Netherlands, 11–14 October 2016; Springer: Cham, Switzerland, 2016; pp. 21–37.
17. Dai, J.; Li, Y.; He, K.; Sun, J. R-fcn: Object detection via region-based fully convolutional networks. In Proceedings of the Advances in Neural Information Processing Systems, Barcelona, Spain, 5–10 December 2016; pp. 379–387.
18. Fu, C.Y.; Liu, W.; Ranga, A.; Tyagi, A.; Berg, A.C. Dssd: Deconvolutional single shot detector. *arXiv* **2017**, arXiv:1701.06659.
19. Zhang, S.; Wen, L.; Bian, X.; Lei, Z.; Li, S.Z. Single-shot refinement neural network for object detection. In Proceedings of the IEEE Conference on Computer Vision and Pattern Recognition, Salt Lake City, UT, USA, 18–23 June 2018; pp. 4203–4212.
20. Redmon, J.; Farhadi, A. YOLO9000: Better, faster, stronger. In Proceedings of the IEEE Conference on Computer Vision and Pattern Recognition, Honolulu, HI, USA, 21–26 July 2017; pp. 7263–7271.
21. Jeong, M.; Ko, B.C.; Nam, J.Y. Early detection of sudden pedestrian crossing for safe driving during summer nights. *IEEE Trans. Circ. Syst. Video* **2016**, *27*, 1368–1380. Available online: https://cvpr.kmu.ac.kr/ (accessed on 28 April 2016). [CrossRef]
22. CVC-09 FIR Sequence Pedestrian Dataset. Available online: http://adas.cvc.uab.es/elektra/enigma-portfolio/item-1/ (accessed on 28 April 2016).
23. Xu, F.; Liu, X.; Fujimura, K. Pedestrian detection and tracking with night vision. *IEEE Trans. Intell. Trans. Syst.* **2005**, *6*, 63–71. [CrossRef]
24. Ko, B.; Kim, D.; Nam, J. Detecting humans using luminance saliency in thermal images. *Opt. Lett.* **2012**, *37*, 4350–4352. [CrossRef] [PubMed]
25. Ge, J.; Luo, Y.; Tei, G. Real-time pedestrian detection and tracking at nighttime for driver-assistance systems. *IEEE Trans. Intell. Trans. Syst.* **2009**, *10*, 283–298.
26. Bertozzi, M.; Broggi, A.; Caraffi, C.; Del Rose, M.; Felisa, M.; Vezzoni, G. Pedestrian detection by means of far-infrared stereo vision. *Comput. Vis. Image Underst.* **2007**, *106*, 194–204. [CrossRef]
27. O'Malley, R.; Jones, E.; Glavin, M. Detection of pedestrians in far-infrared automotive night vision using region-growing and clothing distortion compensation. *Infrared Phys. Technol.* **2010**, *53*, 439–449. [CrossRef]

28. Zhao, X.; He, Z.; Zhang, S.; Liang, D. Robust pedestrian detection in thermal infrared imagery using a shape distribution histogram feature and modified sparse representation classification. *Pattern Recognit.* **2015**, *48*, 1947–1960. [CrossRef]
29. Biswas, S.K.; Milanfar, P. Linear support tensor machine with LSK channels: Pedestrian detection in thermal infrared images. *IEEE Trans. Image Process.* **2017**, *26*, 4229–4242. [CrossRef]
30. Heo, D.; Lee, E.; Ko, B.C. Pedestrian detection at night using deep neural networks and saliency maps. *Electron. Imaging* **2018**, *17*, 1–9. [CrossRef]
31. Cao, Z.; Yang, H.; Zhao, J.; Pan, X.; Zhang, L.; Liu, Z. A new region proposal network for far-infrared pedestrian detection. *IEEE Access* **2019**, *7*, 135023–135030. [CrossRef]
32. Park, J.; Chen, J.; Cho, Y.K.; Kang, D.Y.; Son, B.J. CNN-based person detection using infrared images for night-time intrusion warning systems. *Sensors* **2020**, *20*, 34. [CrossRef]
33. He, K.; Zhang, X.; Ren, S.; Sun, J. Deep residual learning for image recognition. In Proceedings of the IEEE Conference on Computer Vision and Pattern Recognition, Las Vegas, NV, USA, 27–30 June 2016; pp. 770–778.
34. He, K.; Zhang, X.; Ren, S.; Sun, J. Spatial pyramid pooling in deep convolutional networks for visual recognition. *IEEE Trans. Pattern Anal.* **2015**, *37*, 1904–1916. [CrossRef]
35. Redmon, J.; Farhadi, A. Yolov3: An incremental improvement. *arXiv* **2018**, arXiv:1804.02767(1804).
36. Lin, T.Y.; Dollár, P.; Girshick, R.; He, K.; Hariharan, B.; Belongie, S. Feature pyramid networks for object detection. In Proceedings of the IEEE Conference on Computer Vision and Pattern Recognition, Honolulu, HI, USA, 21–26 July 2017; pp. 2117–2125.
37. Wang, X.; Girshick, R.; Gupta, A.; He, K. Non-local neural networks. In Proceedings of the IEEE Conference on Computer Vision and Pattern Recognition, Salt Lake City, UT, USA, 18–23 June 2018; pp. 7794–7803.
38. Li, X.; Zhao, L.; Wei, L.; Yang, M.H.; Wu, F.; Zhuang, Y.; Ling, H.; Wang, J. Deepsaliency: Multi-task deep neural network model for salient object detection. *IEEE Trans. Image Process.* **2016**, *25*, 3919–3930. [CrossRef]
39. Hu, J.; Shen, L.; Sun, G. Squeeze-and-excitation networks. In Proceedings of the IEEE Conference on Computer Vision and Pattern Recognition, Salt Lake City, UT, USA, 18–23 June 2018; pp. 7132–7141.
40. Simonyan, K.; Zisserman, A. Very deep convolutional networks for largescale image recognition. In Proceedings of the International Conference on Learning Representations, San Diego, CA, USA, 7–9 May 2015; pp. 1–14.
41. Erhan, D.; Szegedy, C.; Toshev, A.; Anguelov, D. Scalable object detection using deep neural networks. In Proceedings of the IEEE Conference on Computer Vision and Pattern Recognition, Columbus, OH, USA, 23–28 June 2014; pp. 2147–2154.
42. Russakovsky, O.; Deng, J.; Su, H.; Krause, J.; Satheesh, S.; Ma, S.; Huang, Z.; Karpathy, A.; Khosla, A.; Bernstein, M.; et al. Imagenet large scale visual recognition challenge. *Int. J. Comput. Vis.* **2015**, *115*, 211–252. [CrossRef]
43. Glorot, X.; Bengio, Y. Understanding the difficulty of training deep feedforward neural networks. In Proceedings of the Thirteenth International Conference on Artificial Intelligence and Statistics, Chia Laguna, Sardinia, Italy, 13–15 May 2010; pp. 249–256.
44. Xu, Y.; Xu, D.; Lin, S.; Han, T.X.; Cao, X.; Li, X. Detection of sudden pedestrian crossings for driving assistance systems. *IEEE Trans. Syst. Man Cybern. B* **2012**, *42*, 729–739.
45. Wang, W.; Lai, Q.; Fu, H.; Shen, J.; Ling, H. Salient object detection in the deep learning era: An in-depth survey. *arXiv* **2019**, arXiv:1904.09146.
46. Zhang, S.; Wen, L.; Bian, X.; Lei, Z.; Li, S.Z. Occlusion-aware r-cnn: Detecting pedestrians in a crowd. In Proceedings of the European Conference on Computer Vision, Munich, Germany, 8–14 September 2018; pp. 637–653.
47. Wang, X.; Xiao, T.; Jiang, Y.; Shao, S.; Sun, J.; Shen, C. Repulsion loss: Detecting pedestrians in a crowd. In Proceedings of the IEEE Conference on Computer Vision and Pattern Recognition, Salt Lake City, UT, USA, 18–23 June 2018; pp. 7774–7783.

© 2020 by the authors. Licensee MDPI, Basel, Switzerland. This article is an open access article distributed under the terms and conditions of the Creative Commons Attribution (CC BY) license (http://creativecommons.org/licenses/by/4.0/).

Article

Data-Efficient Domain Adaptation for Semantic Segmentation of Aerial Imagery Using Generative Adversarial Networks

Bilel Benjdira [1,2,*], Adel Ammar [1], Anis Koubaa [1,3] and Kais Ouni [2]

[1] Robotics and Internet of Things Laboratory, College of Computer and Information Sciences, Prince Sultan University, Riyadh 11586, Saudi Arabia; aammar@psu.edu.sa (A.A.); akoubaa@psu.edu.sa (A.K.)
[2] Research Laboratory Smart Electricity & ICT, SEICT, LR18ES44, National Engineering School of Carthage, University of Carthage, Tunis 2035, Tunisia; kais.ouni@enicarthage.rnu.tn
[3] CISTER, INESC-TEC, ISEP, Polytechnic Institute of Porto, 4200-465 Porto, Portugal
* Correspondence: bbenjdira@psu.edu.sa

Received: 17 December 2019; Accepted: 29 January 2020; Published: 6 February 2020

Abstract: Despite the significant advances noted in semantic segmentation of aerial imagery, a considerable limitation is blocking its adoption in real cases. If we test a segmentation model on a new area that is not included in its initial training set, accuracy will decrease remarkably. This is caused by the domain shift between the new targeted domain and the source domain used to train the model. In this paper, we addressed this challenge and proposed a new algorithm that uses Generative Adversarial Networks (GAN) architecture to minimize the domain shift and increase the ability of the model to work on new targeted domains. The proposed GAN architecture contains two GAN networks. The first GAN network converts the chosen image from the target domain into a semantic label. The second GAN network converts this generated semantic label into an image that belongs to the source domain but conserves the semantic map of the target image. This resulting image will be used by the semantic segmentation model to generate a better semantic label of the first chosen image. Our algorithm is tested on the ISPRS semantic segmentation dataset and improved the global accuracy by a margin up to 24% when passing from Potsdam domain to Vaihingen domain. This margin can be increased by addition of other labeled data from the target domain. To minimize the cost of supervision in the translation process, we proposed a methodology to use these labeled data efficiently.

Keywords: deep learning; domain adaptation; semantic segmentation; generative adversarial networks; convolutional neural networks; aerial imagery

1. Introduction

The semantic segmentation task provides for every pixel in the input image a label that defines its semantic class. In remote sensing context, the semantic segmentation of aerial images has an increasing potential for many tasks and applications, like analysis and management of road traffic, monitoring of urban and rural areas, fast interactions in case of emergency, and so on. The growing adoption of Unmanned Aerial Vehicles (UAVs) is behind this increasing potential. High-resolution images can be collected using UAVs from different points of view and processed by the semantic segmentation algorithms to promote the automatic analysis of surveyed scenes.

Since the emergence of Convolutional Neural Networks (CNNs), the area of image analysis algorithms has shown a considerable improvement in accuracy [1–7]. This affected directly the area of semantic segmentation and paved the way towards a variety of CNN-based architectures, like SegNet [8], fully connected network (FCN), PSPNet [9], U-Net [10] and DeepLab [11]. Empirically, if we build a robust dataset and we train on it one of the state-of-the-art algorithms, we will get an accuracy that easily surpasses 80%.

Despite this exciting efficiency, a notable challenge is blocking the use of semantic segmentation algorithms in real cases. In fact, the accuracy of the model will be high only on images belonging to the same domain of the dataset used in training (object representation, resolution, sensor type, lighting conditions). If we test this model in another domain (images collected under conditions different from those of the training set), the accuracy decreases remarkably. This decrease is caused by the domain shift existing between the target and the source domain. Figure 1 shows a real case scenario in which we want to test the semantic segmentation model on a domain different from the source. A considerable shift is remarked between the two domains (location is changed, image coding is changed, resolution is changed).

The straightforward solution to mitigate this intriguing limitation is to train our model on a labeled dataset constructed from the target domain in a supervised or a semi-supervised way [12]. However, this method is costly and time-consuming. To illustrate this, we can note that pixel-labeling of an image from Cityscapes dataset takes nearly 90 minutes on average [13]. The size of the image is 2040 by 1016 pixels. To reduce the labeling time, we can benefit from human crowd intelligence by distributing the labeling task over a set of human crowds [14]. Every set will be allocated to the labeling of one class. This reduces the cost of generating the semantic labeled dataset on the target. Nevertheless, since such solutions are not always immediately available, it is still useful to search for a data-efficient solution of domain adaptation that uses only a minimal set of images for supervision of the domain transfer.

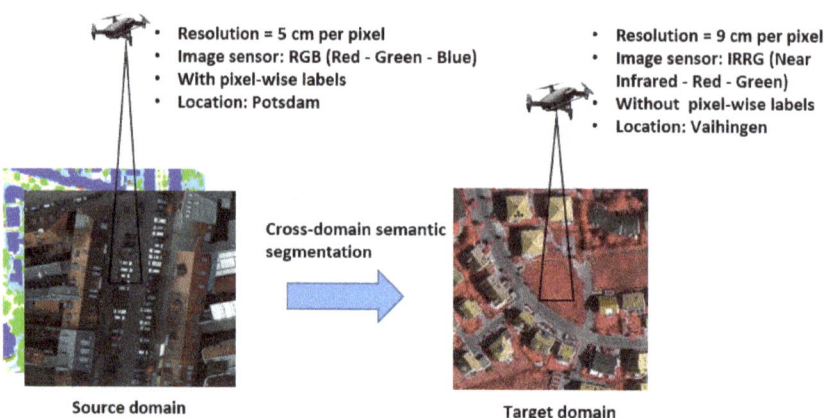

Figure 1. Cross-domains semantic segmentation of aerial imagery.

Domain adaptation is a separate area in machine learning whose objective is to learn how a model performance on a source data distribution can be improved on another target data distribution. Domain adaptation helps in mitigating the domain shift between the target domain data and the source domain data used in training. Typically, a mapping function is designed between the target domain and the source domain. Deep learning models have been used by the recent methods of domain adaptation for the training of such a mapping function [15–18].

Motivated by the current breakthrough made by GANs (Generative Adversarial Networks) [19,20], we developed a data-efficient domain adaptation algorithm based upon an architecture of two GAN networks. Our method aims at handling the case indicated in Figure 1 as well as other cases of the same nature. Our technique is based on converting the image that we want to segment from the target domain to the source domain by passing it through two consecutive GAN networks. The two generative adversarial networks are trained separately. The first GAN network converts the chosen image from the target domain to a semantic segmentation label. The second GAN network converts this semantic label into the source domain. The generated image is proven to conserve the semantic map of the first image and to mimic the source domain characteristics. It will be used as an input to the segmentation model already trained on the source dataset and will generate a better segmentation label, as proven by the experiments we made. The accuracy of final segmentation using our algorithm can be increased by addition of labeled data from the target domain. A methodology is described to use the labeled data efficiently to minimize the cost of supervision in our algorithm. Our work has the following contributions: (1) Our approach is confirmed to mitigate the problem of domain shift for semantic segmentation by a significant margin that can increase with an efficient addition of labeled data from the target domain. (2) The method is validated using ISPRS semantic labeling dataset through the establishment of cross-domain semantic segmentation between Vaihingen and Potsdam datasets. (3) GANs are introduced as a favorable solution for analyzing aerial imagery.

The following is the organization of our paper: Section 2 provides a summary of the related works in the domain adaptation area within semantic segmentation. Section 3 will make an introduction to GANs. Section 4 will describe the different parts of our proposed method. Section 5 will present the experiments made to validate our method and discuss its effectiveness within domain adaptation of semantic segmentation in aerial imagery. Finally, our work is concluded in Section 6, which also deduces the possible extensions of our method.

2. Related Works

In this section, we will discuss the works that treated domain adaptation within semantic segmentation. Generally, it is assumed, in the machine learning context, that the test and the training datasets are belonging to the same distribution. However, there is substantial discordance between them in real cases. This discordance reduces the model's efficiency out of its training domain. Domain adaptation techniques are used mitigate this discordance and to make the model generalizes better over multiple domains.

In computer vision, efforts made on domain adaptation have been more focused on regression and classification tasks [21] not on semantic segmentation. For example, many works treated the training of the models on online images for the classification of objects in reality [22]. Recent works in this area are focusing on improving the adaptability of deep learning algorithms [15,16,23–25].

In semantic segmentation, most of the domain adaptation works are geared towards the use of simulated data [26–31]. In fact, domain adaptation was expected to be used by these works for the improvement of the efficiency of segmentation on real images through the training of the semantic segmentation model on pure synthetic data. FCNs in the wild [32] is one of the earliest works. It uses a pixel-level adversarial loss to guide the model to learn the domain-invariant properties. This aims at making the adversarial classifier not to differentiate between the target and the source domains. The goal is to make its performance equal on the two domains. Hoffman et al. [26] designed CyCADA, which is a model that changes synthetic data (source domain images) into real data style (target domain images) using CycleGAN. The segmentation model is then fed with the converted images to improve its accuracy on dealing with real photos. On the other hand, Zhang et al. [33] pointed out that a curriculum-style learning technique helps in reducing the domain shift. They deduced the target data properties by joining the learning of global label distribution with the learning of local distributions over landmark superpixels. The segmentation network was then trained through regularizing it to follow these features. Sankaranarayanan et al. [34] treated the problem differently by taking the source and the target images as input to an auto-encoder network that processes and regenerates them before feeding them to the segmentation model. CGAN architecture is another approach introduced by Tsai et al. [35] that makes addition of random noise into the source data giving it as input to the segmentation model. They proved that this technique enhances the model's efficiency on the target domains. Huang et al. [36] proposed to train independently two networks for the target and the source domains. The training of the target model is done through its regression into the source model weights since the target domain is without labels. It should be noted that the calculation of an adversarial loss in each layer of both networks is done.

Aerial imagery has its peculiarities that should be considered. Therefore, a dedicated work that takes into consideration its peculiarities should be investigated. Recently, three works targeted the domain adaptation in semantic segmentation of aerial imagery. All of them used an unsupervised approach. The first work is the algorithm proposed by Benjdira et al. [4]. They introduced a GAN architecture to map images in an unsupervised way from the source domain to the target domain without the need for paired data. After that, they translated the source domain dataset into the target domain. Then, the translated dataset is used to fine-tune the segmentation model to enhance its ability to treat images from the target domain. The second work is the work proposed by Tasar et al. [37]. They adopted the same algorithm of Benjdira et al. [4] but changed the GAN architecture into another architecture named ColorMapGAN. This architecture converts the source images into images that the spectral distribution is similar to the target spectral distribution but conserves the semantic information of the source. The third work is introduced by Fang et al. [38]. They made a category-sensitive domain adaptation (CsDA) using a geometry-consistent GAN (GcGAN) embedded into a co-training adversarial learning network (CtALN). Their method is currently the state of the art in unsupervised domain adaptation in semantic segmentation of aerial imagery. Up to our knowledge, no one has treated the domain adaptation in semantic segmentation of aerial imagery using a supervised approach. We aim in this work to target this limitation and to study the implementation of a supervised domain adaptation algorithm based on the concatenation of two GAN networks. We only use a reduced amount of labeled data to guide the model during the training. To demonstrate our algorithm's efficiency, we tested it on the ISPRS semantic segmentation dataset [39]. We performed the domain adaptation for a semantic segmentation network from Potsdam (as source) to Vaihingen (as target).

3. Generative Adversarial Networks (GANs)

3.1. The Generator and the Discriminator

The popularity of GANs has continued to increase because they have many addressable applications. Goodfellow et al. [19] introduced GAN in 2014. A GAN is composed of two separate networks: the generator and the discriminator. During the training process, the generator learns how to generate data that mimics real data, whereas the discriminator learns how to differentiate real data from fake data produced by the generator. The training is operated jointly so that both are competing and playing an adversarial zero-sum game.

During the training, the generator is trying continuously to generate fake data that deceives the discriminator so that it classifies them as real. However, the discriminator is trying continuously to detect the fake data and not to classify them as real. To solve the adversarial zero-sum game during the training, game theory theorems are used. Figure 2 indicates the standard architecture of the GAN network.

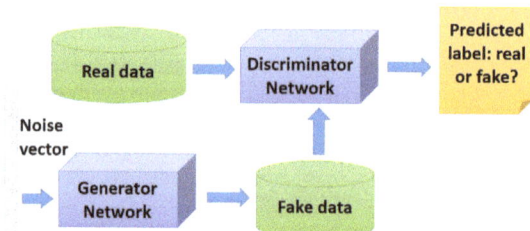

Figure 2. Standard architecture of the GAN.

During training, both networks engage in competition until a Nash equilibrium is reached. In game theory, Nash equilibrium is a status where no player is able improve or deviate his payoff [40].

Equation (1) shows the objective function of GAN:

$$min_G max_D V(Dis, Gen) = \mathbb{E}_{X \sim P_{real\_data}(X)}[log Dis(X)] + \mathbb{E}_{z \sim P_z(z)}[log(1 - Dis(Gen(z)))] \quad (1)$$

where Gen denotes the generator cost function which is trained through the maximization of $Dis(Gen(z))$. On the other hand, Dis represents the discriminator cost function which is trained through the minimization of $Dis(Gen(z))$. While X is the image that is obtained from the distribution of real data p_{real_data}. The noise vector obtained from distribution p_z is denoted by z whereas $Gen(z)$ indicates the fake image that is produced by the generator. Additionally, $\mathbb{E}_{X \sim P_{data}(X)}$ represents the expectation on X, which is acquired by the distribution $P_{real\_data}(X)$. Both Dis and Gen play the two-player minimax game using the value function $V(Gen, Dis)$ [19].

Generative adversarial networks have many applications and implementations [41] with image-to-image translation being the application that may be the most attractive to be used in the domain adaptation's context. The use of GANs in the area of image-to-image translation is further discussed in the following subsection.

3.2. translation from Image-to-Image using GANs

The translation of an image from one domain to another has been the target for many GAN architectures [20,42–44]. Translation of images can be paired [20] or unpaired [45].

3.2.1. Paired Image-to-Image Translation

In this particular area, the GAN model must be trained using labeled pairs of images. With X being the source data, Y being the target data and N being the number of samples in each dataset, each pair of corresponding images $\{x_i\}_{i=0...N}$ and $\{y_i\}_{i=0...N}$ will be used by the model to learn in a supervised way how to translate between X and Y domains. Currently, Pix2pix [20] is the most known model for paired image-to-image translation.

3.2.2. Unpaired Image Translation

GAN is used in unpaired image translation to convert between two sets of images through an unsupervised training. With X being the source data, Y being the target data, N being the number of samples in the source dataset and M the number of samples in the target dataset, $\{x_i\}_{i=0...N}$ and $\{y_i\}_{i=0...M}$ do not correspond with each other and can randomly be obtained from the related domain set. Currently, CycleGAN [45] is the most known model for unpaired image-to-image translation.

4. Proposed Method

4.1. The GAN Architecture

Our method aims at translating images from the target domain to the source domain using two GAN networks as illustrated in Figure 3. The entire procedure makes the target domain images imitate the source domain characteristics, which include the quality of images, types of sensors and resolution, among others. The mapping process between the target and the source is done in two steps. First, the chosen image from the target is mapped into semantic label using the first GAN Network. Then, the generated label is mapped using the second GAN network into another image that looks like images from the source domain distribution. The two networks are trained separately using paired datasets. We used a modified architecture from the standard GAN that is inspired by some recent architectures [20,45].

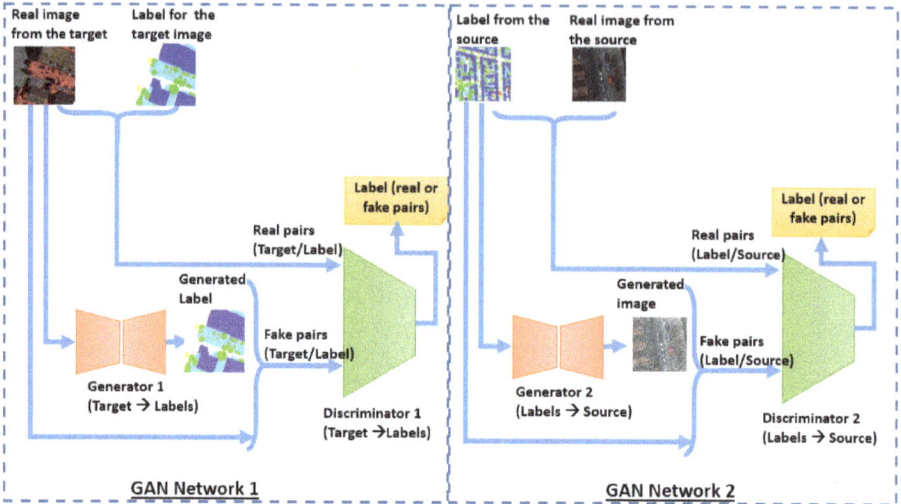

Figure 3. The GAN architecture.

The first GAN Network is designed by substituting the noise vector in the traditional GAN Network by images from the target distribution. The data generated by this GAN Network are the semantic label of the entry image. The training process is done using a small-sized dataset constructed from a few samples from the target domain and their semantic labels. The mapping function of the generator, $G : X \rightarrow L$, learns through an adversarial training how to produce the semantic label L of the target image X. Because the target is supposed to have a minimal set of labels, we assume that the provided labeled images are only a few significant samples from the target dataset. We mean by significant samples the existence of two constraints during the selection of the images. First, the sampled images should contain all the semantic classes of the dataset. Second, for each class, we should choose samples with the most available representations of the class inside the dataset. The discriminator D learns during the adversarial training how to differentiate between real pairs $(X, L_{original})$ and the fake pairs $(X, L_{generated})$. While the discriminator improves its ability to detect fake pairs from real pairs, the generator improves its ability to generate the true semantic labels of the input image. The architecture of the generator and the discriminator in the first GAN network is illustrated in Figure 4.

Concerning the generator, it is an encoder-decoder architecture with skip-connections. It is similar to U-Net [10] architecture. Eight convolutional layers are used for downsampling in the encoder part and similarly, eight deconvolutional layers are used for upsampling in the decoder part. Leaky ReLu [46] is used as the activation layer for all the layers of the encoder except the first one. Leaky ReLu is identical to the standard ReLu in the positive region. However, in the negative part, it has a small slope α. Explicitly, it is defined as $LeakyReLu(x) = x$, if $x >= 0$; and as $LeakyReLu(x) = \alpha x$, if $x < 0$. α has a small value, which gives small gradient where $x < 0$. The encoder part ends by giving us a small feature vector of size $(1 \times 1 \times 512)$. This feature vector will pass through the decoder part to rebuild the original feature vector. We used skip connections to concatenate every layer output in the encoder part with the corresponding layer in the decoder part. ReLu is used as the activation function for all the layers of the decoder. Batch normalization [47] is used after every layer of network except the first layer of the encoder and the last layer of the decoder. We used dropout [48] in the first three layer of the decoder to reduce overfitting. We apply $tanh$ as an activation function to the last layer of the decoder to get the predicted label for the input image.

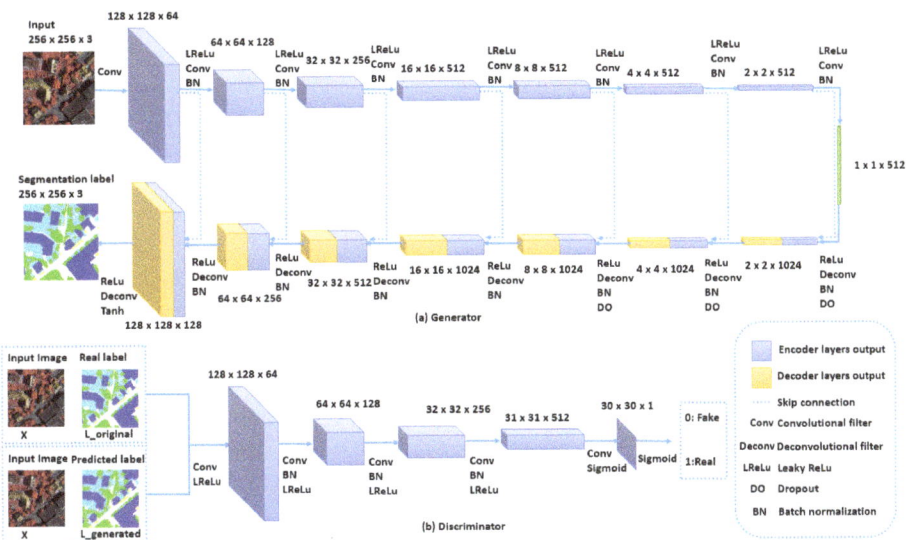

Figure 4. The architecture of the first GAN network: (**a**) the generator and (**b**) the discriminator.

Concerning the discriminator, it takes as input pairs of images from two sets. The first set is images from the target dataset associated with their real label (X_{target}, $L_{original}$). Data from this set should be classified by the discriminator as real pairs of data. The second set is images from the target dataset associated with the generated label from the generator network (X_{target}, $L_{generated}$). Data from this set should be classified by the discriminator as fake pairs of data. In the discriminator, five convolutional layers are used to encode the pair of images (X, L) into a feature vector of size $(30 \times 30 \times 1)$. Then we apply the Sigmoid activation function to this feature vector to get the final binary output of the discriminator {0: fake pairs, 1:real pairs}. As with the encoder part of the generator, we used Leaky ReLu [46] and Batch normalization [47] in every layer except the final one.

Concerning the second GAN Network, it is similar to the first GAN Network. It has a generator and a discriminator as detailed in Figure 3. The architecture of the generator is identical to the architecture of the first generator (See Figure 4) except that it has as input a semantic label and generates an image that imitates images taken from the source dataset. The architecture of the discriminator is also identical to the discriminator of the first GAN (See Figure 4) except that it takes as input pairs of images from two sets. The first set consists of semantic labels associated with their corresponding images from the source dataset (L_{source}, $X_{original}$). The second set consists of semantic labels associated with the generated images from the generator network (L_{source}, $X_{generated}$).

4.2. The Description of the Algorithm

The algorithm adopted to mitigate the domain shift between the source and the target is based on the GAN architecture provided in Figure 3. It is divided into five steps that are illustrated in Figure 5.

There are five steps in the algorithm. Step one begins by the training of a segmentation model on the source domain dataset. The accuracy of the segmentation model could easily get to over 80% if there is a well-structured dataset. In Step 2, we pick out some significant samples from the target domain and label them. The samples are significant if they meet two requirements. First, all the classes should exist in the samples. Second, the most common class representations should be presented. In Step 3, we use these samples to train the first GAN network. Step four makes the training of the second GAN network based on the source dataset provided with labels. Step five is the step where we use our proposed GAN to segment images from the target domain. This step is divided into five sub-steps. In the first sub-step, we pick out any image from the target domain that we want to segment.

Then, we pass it into the generator of the first GAN network to translate it into labels. After that, we pass the generated label into the generator of the second GAN network. This will convert it into an image that imitates images from the source domain. In the fourth sub-step, we pass this generated image into the segmentation network trained in Step 1. We will get then the semantic segmentation map of the image.

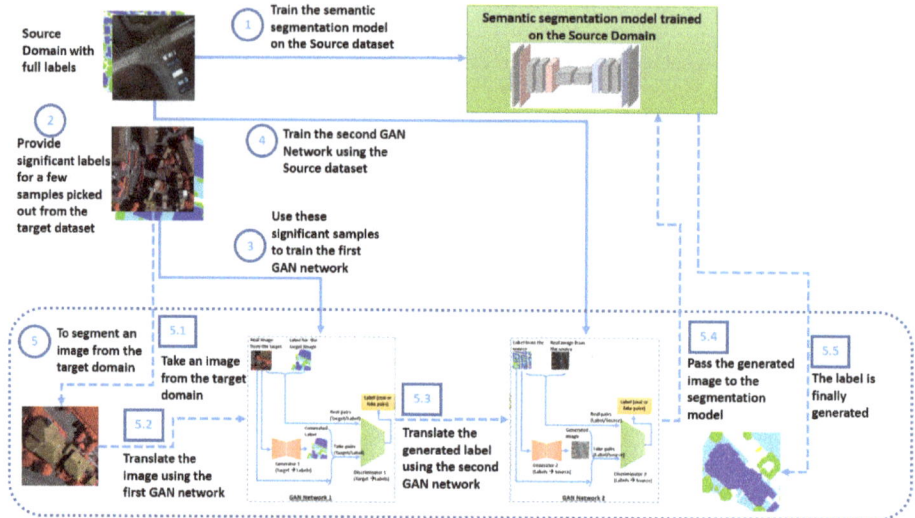

Figure 5. Flowchart of the domain adaptation algorithm.

4.3. Problem Formulation

This section presents the formal modelling of our algorithm. We consider the domain adaptation problem from the target data X_T to the source data X_S. The source data are provided with full semantic labels Y_S while the target data are not initially provided with labels. The first step of the algorithm is concerned with training a semantic segmentation model M_S on the source data. The following equation is correspondent to the source model M_S with the use of the cross-entropy loss:

$$L_{M_S}(M_S, X_S, Y_S) = -\mathbb{E}_{(x_s,y_s) \sim (X_S,Y_S)} \sum_{c=1}^{C} \mathbb{1}_{[c=y_s]} log(Softmax(M_S^{(c)}(x_s))) \qquad (2)$$

where $\mathbb{E}_{(x,ys) \sim (X_S,Y_S)}$ denotes the expectation on x_s, y_s, which is drawn by the X_S and Y_S distribution. C is the number of classes of the segmentation task. $\mathbb{1}_{[c=y_s]}$ represents the loss for the class c separated from other classes. Due to the advancements in the semantic segmentation field, M_S could have a good accuracy if the source data are well constructed. However, when we use this model to segment images from the target domain, the accuracy will decrease because of the domain shift that exists between the source and the target. Hence, we will apply our proposed algorithm and begin by picking out significant samples from the target domain and labeling them. This small dataset will guide the mapping process from the target to the source. It will be used to train the first GAN Network, which will learn through an adversarial loss how to generate pixel-wise labels of the target images.

The mapping model from the target domain to the semantic labels $G_{T \rightarrow L}$ is trained for a segmentation task in an adversarial manner. The discriminator $D_{T \rightarrow L}$ will be trained on the other side to detect fake pairs of data from the real ones. This GAN model can be seen as a standard GAN where we applied three modifications. First, we substituted the noise vector by input images from the target domain. Second modification is to set the output of the GAN to semantic segmentation labels of the input image. Third, is to set the input for the discriminator as pair of image for the target domain x_t, and semantic label real or generated by the discriminator (l_{t_real} or $l_{t\_generated}$).

Also, the GAN model can be treated as a modification of a conditional GAN [49]. Generally, conditional GANs or cGANs make a mapping between an input and an output. The input are formed by an image x from a distribution X concatenated to a noise vector z. The output is an image y from a distribution Y. The mapping is formulated by $G_{cGAN} : \{x,z\} \longrightarrow y$ and the loss function of a conditional GAN can be formulated as:

$$L_{cGAN}(G,D,X,Y,z) = \mathbb{E}_{(x \sim X, y \sim Y)}[\log D(x,y)] + \mathbb{E}_{(x \sim X, z)}[\log(1 - D(x, G(x,z)))] \quad (3)$$

During the training, G (the generator) is trying to minimize this loss, whereas D (the discriminator) is trying to maximize it. The loss function can be expressed more clearly as:

$$L = arg\ min_G\ max_D\ L_{cGAN}(G,D,X,Y,z) \quad (4)$$

As proven in [20], mixing this GAN objective with L1 loss gives beneficial results. In fact, this helps the generator to be more able to fool the discriminator by being closer to the ground truth in an L1 sense. This loss does not affect the discriminator. The L1 loss related to the cGAN can be expressed as:

$$L_{L1} = \mathbb{E}_{(x \sim X, y \sim Y, z)}[\|y - G(x,z)\|_1] \quad (5)$$

The final objective of the cGAN can be expressed as:

$$L = \lambda_1 arg\ min_G\ max_D\ L_{cGAN}(G,D,X,Y,z) + \lambda_2 L_{L1} \quad (6)$$

where λ_1 and λ_2 are the weights for the original GAN loss and the L1 loss, respectively. Concerning the noise vector z, it is added to the cGAN to give stochasticity inside the given output. By removing it and passing only the input x_s, the generator tends more to give deterministic outputs and fails to grasp the whole entropy of the distribution they want to learn. Normally, Gaussian noise is used for the vector z. In the experiments, this noise vector does not prove high efficiency for capturing the high data variability. The model learns during the training to ignore the noise. Therefore, we did not use the noise vector. We used, instead, the dropout technique in the three first layers of the decoder part of the generator. The dropout has the effect to add some minor stochasticity in the generated output. Hence, the loss function of the first GAN network we implemented is:

$$Loss(GAN1) = \lambda_1 arg\ min_G\ max_D [\mathbb{E}_{(x_t \sim X_T, l_t \sim L_T)}[\log D(x_t, y_t)] + \\ \mathbb{E}_{(x_t \sim X_T)}[\log(1 - D(x_t, G(x_t)))] + \lambda_2 \mathbb{E}_{(x_t \sim X_T, l_t \sim L_T)}[\|y_t - G(x_t)\|_1] \quad (7)$$

where x_t are the input images sampled from the target distribution data X_T, l_t is the label associated with this input image and sampled from the target label distribution data L_T.

Similarly, the loss function of the second GAN network we implemented is defined as:

$$Loss(GAN2) = \lambda_1 arg\ min_G\ max_D [\mathbb{E}_{(l_s \sim L_S, x_s \sim X_S)}[\log D(l_s, x_s)] + \\ \mathbb{E}_{(l_s \sim L_S)}[\log(1 - D(l_s, G(l_s)))] + \lambda_2 \mathbb{E}_{(l_s \sim L_S, x_s \sim X_S)}[\|x_s - G(l-s)\|_1] \quad (8)$$

where x_s are the images sampled from the source distribution data X_S, l_s is the label associated with this image and sampled from the source label distribution data L_S. $GAN2$ is mapping from the semantic label to the source images.

To perform a segmentation of an image from the target domain X_T, we translate it to source domain X_S in two steps. Step one makes the translation from the target domain to semantic label domain using the generator of $GAN1$. Step 2 makes the translation from the semantic label domain to the source domain using the generator of $GAN2$. The source semantic segmentation model M_S will be more able to segment the final translated image as it belongs more to the source domain on which it was trained. This will be more emphasized in the Experimental Section.

5. Experimental Results

This section aims at confirming the efficiency of our algorithm by describing the experiments that were implemented as well as discussing the findings.

5.1. The Datasets and the Evaluation Metrics

5.1.1. The Datasets

ISPRS (WGII/4) 2D semantic segmentation benchmark dataset [39] was used for validating our methodology. This dataset is provided by the ISPRS 2D semantic labeling challenge which provides a whole platform for the evaluation of semantic segmentation algorithms in aerial imagery context. We used Potsdam and Vaihingen datasets that are freely accessible by the community. Every image is provided with DSM (digital surface model) data. However, we used only the image data because we are targeting the domain adaptation using image data only. The two datasets comprise of very high-resolution photos with a 5 cm per pixel for Potsdam photos and 9 cm per pixel for Vaihingen photos. This difference of resolution represents one of the domain shift factors between the two domains. The VHR (Very High Resolution) helps in minimizing the interclass variance and maximizing the intraclass variance through the provision of more details about the objects represented in the images.

Every image in the two datasets is afforded with its semantic segmentation label that is categorized into six classes of ground objects: car, low vegetation, building, tree, clutter/background, and impervious surfaces. Clutter/background includes any ground object excluded from the five classes, while impervious surfaces show a paved area without any structure on it. In the Vaihingen dataset, there are 33 TOP images whose sizes are around 2000 × 2000 pixels. There are three channels in the TOP file: green, red, and the infrared bands. Out of the 33 TOP images, 27 were used for the training, and the remaining six were used for test. The Potsdam dataset contains 38 TOP images of size 6000 × 6000 pixels. These TOP files have three spectral channels: blue, green, and red. We subdivided these images into 32 TOP images for train, and the remaining 6 for test. For training of the segmentation model, the images were divided to squares of size 512 × 512 pixels used to feed the network. Samples from Vaihingen and Potsdam ISPRS datasets are shown in Figure 6.

Pixel distribution is not balanced across the six classes. Some classes like Buildings are more redundant than other classes like cars. Each class percentage in proportion to the whole number of pixels in the dataset is represented in Table 1.

Table 1. The distribution of pixels among categories.

Category	Potsdam	Vaihingen
Buildings	28.2%	26.9%
Impervious Surfaces	29.9%	29.3%
Low vegetation	20.9%	19.4%
Trees	14.4%	22.4%
Clutter	4.8%	0.7%
Cars	1.7%	1.3%

Figure 6. Image samples from Vaihingen and Potsdam ISPRS datasets.

5.1.2. The Analysis of the Domain Shift Factors

Three factors are responsible for the domain shift between Potsdam (the source domain), and Vaihingen (the target domain): Sensor variation, the variation of class representation, and the variation of resolution. Beginning by the sensor variation factor, Potsdam images are captured using RGB (Red-Green-Blue) sensor while Vaihingen images are captured using IRRG (Infrared, Red and Green) sensor. For example, green color is transformed to red in Vaihingen images. This makes the model which is trained on Potsdam images to fail in recognizing classes normally associated with green color like trees and low vegetation. Concerning the variation of resolution, we note that Vaihingen images are captured by using 9 cm per pixel resolution while those of Potsdam are captured using 5 cm per pixel resolution. This variation affects the ability of the model to recognize classes that are trained on a specific scale of resolution (like cars for example). Passing to the third factor, which is the difference of class representation, it is the most delicate factor to treat. To illustrate this domain shift factor, we can take the case of low vegetation class. Low vegetation in Potsdam are mostly grass areas in the modern town style. In Vaihingen, low vegetation class has different patterns and representations. In fact, they correspond to agricultural zones containing different types of vegetation. This difference of patterns on the same class affects the model ability when passing from a domain to another. We made a careful analysis of the domain between Potsdam and Vaihingen images for the six classes and the results are summarized in Table 2. This table helps us in studying the effect of our algorithm on every domain shift factor.

Table 2. Domain shift analysis when passing from Potsdam to Vaihingen.

Domain Shift Factor	Resolution	Sensor	Class Representation
Trees	low	high	medium
Cars	low	low	low
Clutter	low	high	high
Impervious Surfaces	low	low	low
Buildings	low	high	low
Low vegetation	low	high	high

5.1.3. Evaluation Metrics

To evaluate the semantic segmentation algorithms, four metrics are used: the accuracy, the recall, the precision and the F1 score. These metrics are calculated using TN (True Negatives), TP (True Positives), FN (False Negatives) and FP (False Positives). Considering a semantic segmentation class C, TP is the number of pixels of class C classified successfully as C. TN corresponds to the number of pixels that are not C and the algorithm did not classify them as C. FN corresponds to the number of pixels that belong to the class C but the algorithm did not classify them as C. FP is the count of pixels that are classified incorrectly as C while they do not belong really to C. These metrics are expressed below:

$$Accuracy = \frac{TP + TN}{TP + TN + FP + FN} \tag{9}$$

$$Precision = \frac{TP}{TP + FP} \tag{10}$$

$$Recall = Sensitivity = \frac{TP}{TP + FN} \tag{11}$$

$$F1 = Dice = \frac{2 \times TP}{2 \times TP + FP + FN} \tag{12}$$

We used also a fifth metric, which is the Intersection over Union (IoU) to evaluate the global segmentation efficiency. IoU is computed for every class separately before deducing the mean IoU for all the classes. Below is the expression of the IoU for two different sets of data A and B:

$$IoU(A, B) = \frac{size(A \cap B)}{size(A \cup B)} = \frac{TP}{TP + FP + FN} \tag{13}$$

5.2. Experimental Settings

5.2.1. Step 1: Training of the Semantic Segmentation Model

The algorithm begins by training the semantic segmentation model on the source dataset. Concerning the selection of the source dataset, we demonstrated in Section 5.2.6 (Discussion part) that our algorithm shows higher efficiency if we choose Potsdam domain as source and Vaihingen domain as target. In fact, Potsdam is far larger than Vaihingen (3800 images in Potsdam compared to 459 images in Vaihingen). Hence, to maximize the efficiency of our algorithm, we should select a large dataset as the source to learn the global patterns of the data. Then, our domain adaptation algorithm is used to treat domain shift that exists between the source and the target datasets. Consequently, we selected in our experiments Potsdam dataset as the source dataset and Vaihingen as the target dataset. Once the dataset was ready, we applied a state-of-the-art semantic segmentation model adapted to our requirements in aerial imagery. We chose the Bilateral Segmentation Network (BiSeNet) [50], which is the fastest segmentation model tested on the Cityscapes dataset [13]. It achieves a speed of 65.5 frames per second with a mean IoU of 74.7 % on this dataset [51]. In the processing of aerial images, the speed factor is very important to have the ability to process video streams captured in real time from a drone. The architecture of BiSeNet is represented in Figure 7.

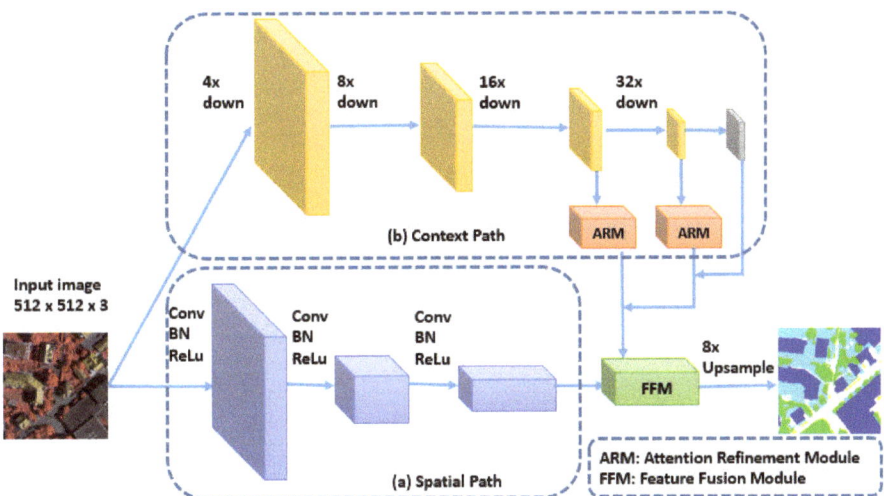

Figure 7. The Architecture of BiSeNet (Bilateral Segmentation Network).

A GPU machine containing the following features was used to carry out experiments associated with this study:

- Processor: Intel Core i9 (Coffee Lake architecture, 6 cores)
- GPU: Nvidia GTX 1080, 8GB dedicated
- Memory: 32 GB RAM
- Operating system: Windows 10 and Linux (Ubuntu 16.04)

To train BiSeNet on the Potsdam dataset, we used semantic segmentation Suite [52], which is an open-source framework where several segmentation models are implemented in Tensorflow [53]. ResNet101 [54] was used as the front end for the BiSeNet network. The training is run for 80 epochs, we set 1 image as the batch size. Image augmentation techniques are not used. ADAM [55] is used as an optimizer during the training with a learning rate 0.0001. As shown in Figure 8, the average segmentation accuracy surpasses 85% on the validation dataset in less than 15 epochs.

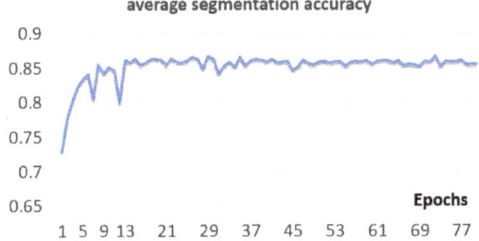

Figure 8. Progress of segmentation accuracy of BiseNet trained on Potsdam.

The Figure 9 shows the convergence of the loss of BiSeNet over the epochs.

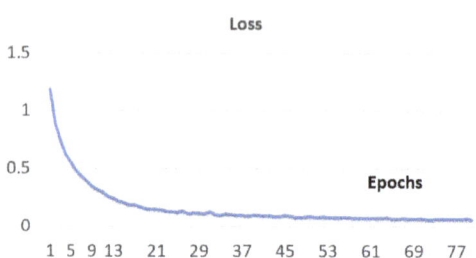

Figure 9. Curve of the loss during the training of BiseNet on Potsdam dataset.

After the training, model weights are saved to be used later in Step 5.

5.2.2. Step 2: Label Significant Data Samples from the Target Domain

To use data efficiently from the target domain, Step 2 consists of picking out significant samples from the target domain and semantically labeling them. The samples are significant if they respect two conditions. The first condition is that they contain pixels from all the classes on which the system is trained. Preferably, the class representation should be as balanced as possible, but this is not necessary. Sometimes, assuring balanced pixel distribution between classes on real aerial images is not possible. The second condition that should be met by the selected samples is that for every class, we should provide the most common patterns in the target domain. Once the samples are chosen, we semantically label them in respect to our targeted classes.

5.2.3. Step 3: Training the First GAN Network

We began by training the first GAN network using the set of labeled data we made in Step 2. We trained the GAN network many times using different sizes of sampled data to study the effect of the number of sampled images on increasing the accuracy of the segmentation. We tried 7 sizes of sampled images from the target domain (Vaihingen): 1, 3, 12, 23, 47, 94, 188. All the selected images are of size 512 ∗ 512. The GAN network learns during the training how to translate images from the target domain to the semantic label domain. The GAN architecture was implemented using Tensorflow [53]. We set the value 0.2 for the slope α of Leaky ReLu. We used ADAM optimizer with an initial learning rate of 0.0002 and momentum term β 0.5. The weight for the L_1 loss is set to 100 and the weight for the GAN weight is set to 1.

5.2.4. Step 4: Training the Second GAN Network

We passed to the fourth step of the algorithm, which is the training of the second GAN network that maps from the label domain to the source domain (Potsdam). We used the full provided source target for this training (3800 images of size 512 × 512). The GAN is implemented in Tensorflow [53]. We used the value 0.2 for the slope of Leaky ReLu. We used ADAM optimizer using an initial learning rate set to 0.0002 and a momentum term β set to 0.5. We run the training until convergence of the loss of the discriminator and the generator. Figure 10 shows some images generated in the source domain (Potsdam) from the semantic label. The generated images are mimicking the characteristics of real images from the source domain.

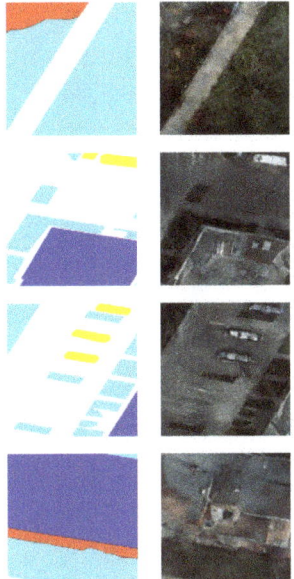

Figure 10. Translation of images from the label domain to the source domain using the second GAN network.

5.2.5. Step 5: Segment Images from the Target Domain

Once we finished the training of the first and the second GAN network, we apply Step 5 for segmentation of images from the target domain. We began by selecting the image we want to segment from the target domain (Vaihingen). Then we apply the first GAN network to translate this image to the label domain. The corresponding semantic label image will be passed into the second GAN network to generate the corresponding image in the source domain (Potsdam). The semantic segmentation model already trained on the source domain will be more able to segment this generated image because it is in the same domain it was trained on.

5.2.6. Discussion

The results of segmentation are always better using the generated image than the original one. Figure 11 shows the result of segmentation of some selected images from the target domain before and after application of our proposed algorithm. We can see clearly that our algorithm improves the quality of segmentation without needing too much data. The results in Figure 11 are generated using first GAN network trained only on 23 labeled images of size $512 * 512$ from the target domain.

The improvement in segmentation accuracy increases with the number of labeled images picked out from the target domain. We tested different trainings of the GAN 1 model using 1, 3, 12, 23, 47, 94 and 188 labeled images from the target domain.

To select the source dataset from Vaihingen and Potsdam, we applied our algorithm twice. The first is done using Potsdam as source and Vaihingen as target, results are shown in Table 3. The second is done using Vaihingen as source and Potsdam as target, results are shown in Table 4. Both tables show the improvement made by the algorithm in segmentation accuracy, precision, sensitivity, dice coefficient and IoU for the different numbers of sampled images from the target domain.

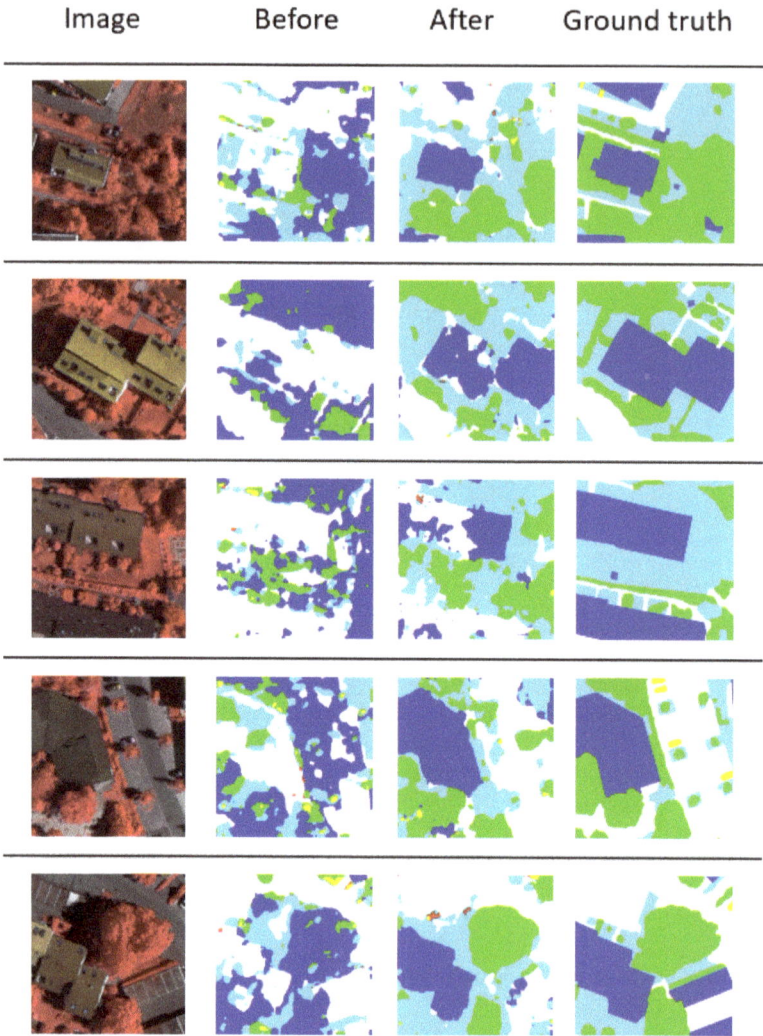

Figure 11. Segmentation of images from the target domain before and after application of our proposed algorithm.

Table 3. Segmentation Accuracy, Precision, Sensitivity, Dice Coefficient and IoU score for different numbers of sampled images from the target domain (Potsdam as source and Vaihingen as target).

Number of Images from Target	Average Accuracy	Precision	Sensitivity	Dice Coef	IoU
0	0.345	0.345	0.345	0.316	0.175
1	0.368	0.445	0.368	0.360	0.176
3	0.422	0.470	0.422	0.404	0.214
12	0.474	0.513	0.474	0.455	0.253
23	0.507	0.541	0.507	0.488	0.281
47	0.524	0.559	0.524	0.505	0.297
94	0.559	0.591	0.559	0.543	0.327
188	0.588	0.625	0.588	0.572	0.349

Table 4. Segmentation Accuracy, Precision, Sensitivity, Dice Coefficient and IoU score for different numbers of sampled images from the target domain (Vaihingen as source and Potsdam as target).

Number of Images from Target	Average Accuracy	Precision	Sensitivity	Dice Coef	IoU
0	0.334	0.335	0.334	0.288	0.169
192	0.363	0.365	0.363	0.318	0.179

As illustrated in Tables 3 and 4, the algorithm efficiency is far better if we select Potsdam as the source. For similar number of labeled images (188 images in first case and 192 images in second case), improvement in average accuracy is 0.243 in first case and 0.026 in the second case. In fact, Potsdam dataset is far larger than Vaihingen dataset (3800 images in Potsdam and 459 images in Vaihingen). Training the segmentation model on a large dataset helps to better capture the global patterns of the data. Then, our algorithm will be more efficient in minimizing the domain shift that exists between the source and the target. This can be seen when selecting Potsdam as source in Table 3. If the source dataset is small, the segmentation model will be less able to learn the global patterns of the data. In this case, our algorithm will not be able to treat efficiently the domain shift. This can be seen when selecting Vaihingen as source a in Table 4. Hence, our algorithm works more efficiently if the source dataset is large enough to make the segmentation model capture the global patterns of the data.

As shown in Table 3, without our algorithm, the total accuracy was 34.5%. Using only 1 labeled images, accuracy increases to 36.8%. The more labeled images are added, the more the accuracy increases until reaching 58.8% for 188 labeled images from the target domain. 188 images represents only 4.9% of the images that was needed to train the segmentation model on the source dataset (3800 images of size 512 × 512 which proves the data efficiency of our algorithm. Figures 12–16 respectively display the curves of accuracy, precision, sensitivity, F1 score and IoU for different numbers of sampled images from the target domain.

We deduced from the above curves that our algorithm increased the segmentation metrics by a significant margin, which explains the visible amelioration of the segmented map shown in Figure 11.

To judge the global efficiency of our algorithm, we compared it with two other domain adaptation algorithms. The first is FCNs in the wild [32]. It is a domain adaptation algorithm applied in general semantic segmentation. The second is the unsupervised algorithm introduced by Benjdira et al. [4]. It was designed specifically for domain adaptation in semantic segmentation of aerial imagery. We provided in Table 5 the comparison of average accuracy, dice coefficient (F1 score) and IoU in the task of domain adaptation from Potsdam to Vaihingen.

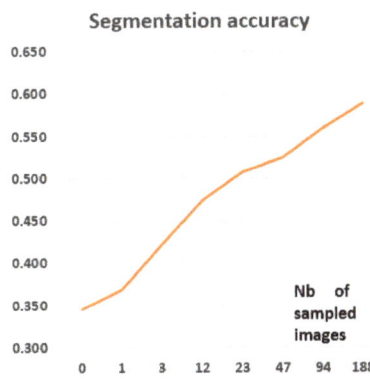

Figure 12. Segmentation accuracy for different numbers of sampled images from the target domain.

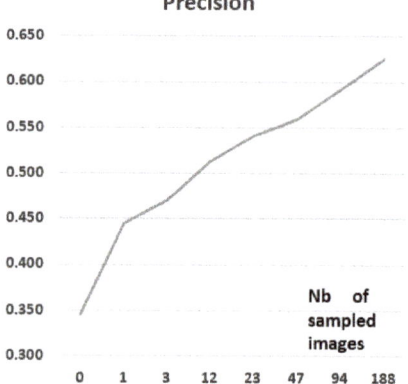

Figure 13. Segmentation precision for different numbers of sampled images from the target domain.

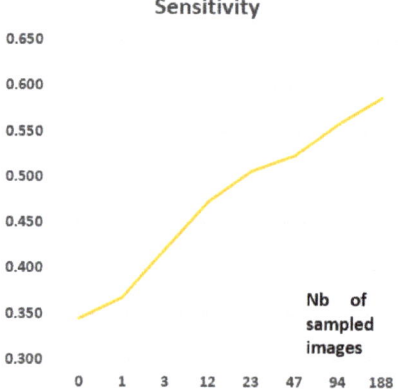

Figure 14. Segmentation sensitivity for different numbers of sampled images from the target domain.

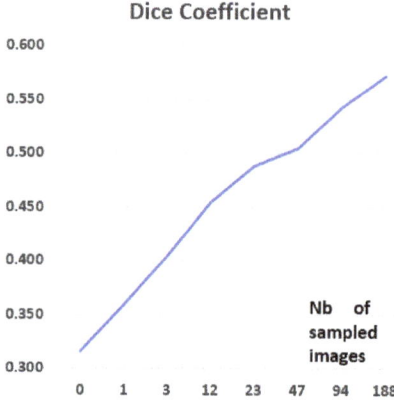

Figure 15. Segmentation dice coefficient for different numbers of sampled images from the target domain.

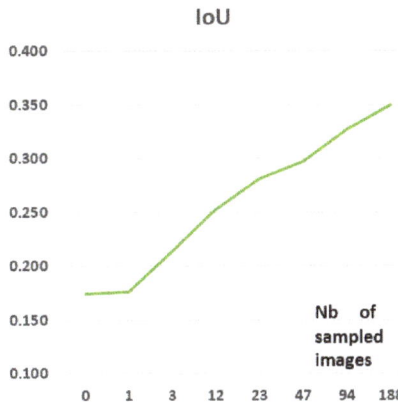

Figure 16. Segmentation IoU for different numbers of sampled from the target domain.

Table 5. Segmentation Accuracy, Dice Coefficient and IoU score for different domain adaptation algorithms (Potsdam as source and Vaihingen as target).

Method	Average Accuracy	Dice Coef	IoU
Without Domain Adaptation	0.345	0.316	0.175
FCNs in the wild	0.486	0.413	0.309
Unsupervised Method in [4]	0.520	0.490	0.300
Ours (188 images)	0.588	0.572	0.349

As shown in Table 5, our algorithm outperforms the other two algorithms in all measures. As a conclusion, the final user can choose between the unsupervised approach introduced by Benjdira et al. [4] and the current supervised approach. If the user does not want to invest time in labeling and wants a medium accuracy, he can choose the unsupervised approach. If he cares most about the accuracy, he can use the supervised approach presented in this paper. The choice will be a tradeoff between the labeling cost and the accuracy of domain adaptation.

Going deeper in the analysis of our algorithm, we studied its effect on the improvement of the segmentation accuracy for every class apart. Table 6 shows the improvement noted in each class for different numbers of sampled images. Moreover, we can validate that for every class, the more the size of labeled images increases, the more the segmentation accuracy is improved. This has some limited exceptions with some local decrease in the accuracy but did not affect the global improvement behavior. Only one class, the clutter background, knows a small decrease in accuracy and this is because there is no specific pattern to be learned for this class. It represents all the components that cannot be included within the other five classes, and the small number of samples is not sufficient for the first GAN network to capture the different patterns existing for this class. On the other side, classes Trees, Low vegetation and Building know a significant increase of accuracy by a margin of 47.7%, 21.9% and 31.6%, respectively. This is because the picked samples from the target domain were sufficient for the first GAN network to learn most of the patterns that exist for the class.

Table 6. Improvement of per-class accuracy for different numbers of sampled images from the target domain.

Nb of Images	Imp. Surf.	Building	Low Veget.	Tree	Car	Clutter Backgr.
0	0.583	0.227	0.383	0.062	0.400	0.935
1	0.591	0.265	0.368	0.311	0.323	0.893
3	0.477	0.303	0.553	0.417	0.323	0.893
12	0.538	0.439	0.559	0.402	0.325	0.894
23	0.602	0.467	0.573	0.421	0.338	0.894
47	0.605	0.416	0.575	0.520	0.385	0.893
94	0.634	0.463	0.618	0.509	0.405	0.893
188	0.685	0.543	0.602	0.539	0.408	0.893

Figure 17 shows the curves of improvement of segmentation accuracy for every class for different sizes of the labeled images from the target domain.

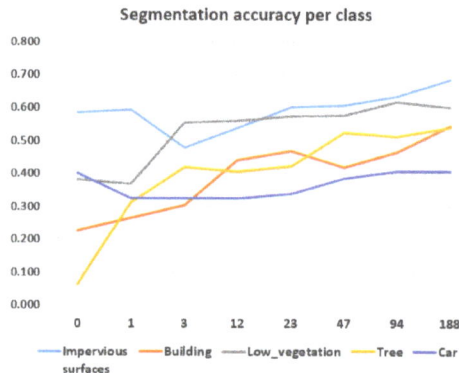

Figure 17. Segmentation accuracy per class for different sizes of the labeled images from the target domain.

Concerning the domain shift analysis of our algorithm, we note that it has no practical effect on classes that are not subject to domain shift (like class Impervious Surfaces and class Cars). In fact, our algorithm made an improvement of 0.8% for the class Car and 10.2% for the class Impervious Surfaces. On the other side, if the sensor domain shift factor is high on the class (See Table 2), the algorithm improves remarkably the segmentation accuracy. The improvement becomes smaller if it is combined with a high effect of other factors. For example, the class Trees and Buildings have a high effect of sensor variation factor and respectively medium and low effects of the class representation factor. The improvement is then 47.7% and 31.6%, respectively. We conclude, then, the efficiency of our algorithm in treating these cases of domain shift factors. On the other side, classes Low vegetation and clutter have a high effect of the sensor variation factor and a high effect of the class representation factor (the effect is much higher in the clutter class). The improvement is respectively 21.9% and −4.2%. We can conclude that the high effect of other domain shift factors reduces the impact of our algorithm, however, small to medium domain shift factors could be mitigated due to the supervision made by the labeled images. If the class patterns are presented in the labeled images (example: class Tree), the domain shift could be mitigated at a good degree. If the class patterns are so diversified and could not be provided in the small set of labeled images (example: class clutter background), the domain shift will be mitigated proportionally. This can be considered to be a limitation to our data-efficient approach. Hence, our algorithm is very efficient in treating the sensor variation factor. Concerning the class representation factor, our algorithm will be efficient in cases the effect is low to medium and most of the class patterns are provided within the labeled dataset.

6. Conclusions

In this study, we proposed a data-efficient supervised approach for domain adaptation in the context of semantic segmentation of aerial imagery. The algorithm is based on two GAN networks to translate images from the target to the source domain. This translation needs the provision of a small set of labeled images from the target domain. The method improves the segmentation accuracy by a margin up to 24% providing only 4.9% of the labeled data needed to train the segmentation model on the source dataset. Our method is confirmed to be efficient in treating the domain shift resulting from the sensor variation factor. It is also efficient in treating low to medium degrees of class representation factor if most of the class patterns are provided within the labeled data. Our work confirms the potential of GANs in aerial imagery analysis. Nevertheless, our algorithm should be combined with a solution to treat cases where we have high degree of class representation factor. Also, we should provide a remedy for cases where the class patterns are so diversified that they cannot be provided in the sampled images from the target domain. This can be solved using a semi-supervised approach to alleviate the manual work needed to label data.

Author Contributions: B.B. and A.A. designed the method. A.A. implemented the method. B.B. wrote the paper. A.K. and K.O. contributed to the supervision of the work. All authors have read and agreed to the published version of the manuscript.

Funding: This research was funded by Prince Sultan University.

Acknowledgments: This work is supported by Robotics and Internet of Things Lab (RIOTU), Prince Sultan University.

Conflicts of Interest: The authors declare no conflict of interest.

References

1. Alhichri, H.; Jdira, B.B.; Alajlan, N. Multiple Object Scene Description for the Visually Impaired Using Pre-trained Convolutional Neural Networks. In Proceedings of the International Conference on Image Analysis and Recognition, Póvoa de Varzim, Portugal, 13–15 July 2016; Springer: Berlin/Heidelberg, Germany, 2016; pp. 290–295.
2. Benjdira, B.; Khursheed, T.; Koubaa, A.; Ammar, A.; Ouni, K. Car Detection using Unmanned Aerial Vehicles: Comparison between Faster R-CNN and YOLOv3. In Proceedings of the 2019 1st International Conference on Unmanned Vehicle Systems-Oman (UVS), Muscat, Oman, 5–7 February 2019; IEEE: Piscataway, NJ, USA, 2019; pp. 1–6.
3. Al Rahhal, M.M.; Bazi, Y.; Al Zuair, M.; Othman, E.; BenJdira, B. Convolutional neural networks for electrocardiogram classification. *J. Med Biol. Eng.* **2018**, *38*, 1014–1025. [CrossRef]
4. Benjdira, B.; Bazi, Y.; Koubaa, A.; Ouni, K. Unsupervised Domain Adaptation Using Generative Adversarial Networks for Semantic Segmentation of Aerial Images. *Remote. Sens.* **2019**, *11*, 1369. [CrossRef]
5. Ammour, N.; Alhichri, H.; Bazi, Y.; Benjdira, B.; Alajlan, N.; Zuair, M. Deep learning approach for car detection in UAV imagery. *Remote. Sens.* **2017**, *9*, 312. [CrossRef]
6. Singh, V.K.; Rashwan, H.A.; Romani, S.; Akram, F.; Pandey, N.; Sarker, M.M.K.; Saleh, A.; Arenas, M.; Arquez, M.; Puig, D.; et al. Breast tumor segmentation and shape classification in mammograms using generative adversarial and convolutional neural network. *Expert Syst. Appl.* **2019**, *139*, 112855. [CrossRef]
7. Ammar, A.; Koubaa, A.; Ahmed, M.; Saad, A. Aerial Images Processing for Car Detection using Convolutional Neural Networks: Comparison between Faster R-CNN and YoloV3. *arXiv* **2019**, arXiv:1910.07234.
8. Badrinarayanan, V.; Kendall, A.; Cipolla, R. SegNet: A Deep Convolutional Encoder-Decoder Architecture for Image Segmentation. *IEEE Trans. Pattern Anal. Mach. Intell.* **2017**, *39*, 2481–2495. [CrossRef]
9. Zhao, H.; Shi, J.; Qi, X.; Wang, X.; Jia, J. Pyramid Scene Parsing Network. In Proceedings of the 2017 IEEE Conference on Computer Vision and Pattern Recognition (CVPR), Honolulu, HI, USA, 21–26 July 2017. [CrossRef]

10. Ronneberger, O.; Fischer, P.; Brox, T. U-net: Convolutional networks for biomedical image segmentation. In Proceedings of the International Conference on Medical Image Computing and Computer-Assisted Intervention, Munich, Germany, 5–9 October 2015; Springer: Berlin/Heidelberg, Germany, 2015; pp. 234–241.
11. Chen, L.C.; Papandreou, G.; Kokkinos, I.; Murphy, K.; Yuille, A.L. Deeplab: Semantic image segmentation with deep convolutional nets, atrous convolution, and fully connected crfs. *IEEE Trans. Pattern Anal. Mach. Intell.* **2018**, *40*, 834–848. [CrossRef]
12. Dong, L.Y.; Sui, P.; Sun, P.; Li, Y.L. Novel naive Bayes classification algorithm based on semi-supervised learning. *Jilin Daxue Xuebao (Gongxueban)/J. Jilin Univ. (Eng. Technol. Ed.* **2016**, *46*, 884–889. [CrossRef]
13. Cordts, M.; Omran, M.; Ramos, S.; Rehfeld, T.; Enzweiler, M.; Benenson, R.; Franke, U.; Roth, S.; Schiele, B. The cityscapes dataset for semantic urban scene understanding. In Proceedings of the IEEE conference on computer vision and pattern recognition, Las Vegas, NV, USA, 27–30 June 2016; pp. 3213–3223.
14. Sun, P.; Brown, C.; Beschastnikh, I.; Stolee, K.T. Mining Specifications from Documentation using a Crowd. In Proceedings of the 2019 IEEE 26th International Conference on Software Analysis, Evolution and Reengineering (SANER), Hangzhou, China, 24–27 February 2019; pp. 275–286.
15. Tzeng, E.; Hoffman, J.; Darrell, T.; Saenko, K. Simultaneous deep transfer across domains and tasks. In Proceedings of the IEEE International Conference on Computer Vision, Santiago, Chile, 7–13 December 2015; pp. 4068–4076.
16. Long, M.; Cao, Y.; Wang, J.; Jordan, M.I. Learning transferable features with deep adaptation networks. *arXiv* **2015**, arXiv:1502.02791.
17. Tzeng, E.; Hoffman, J.; Saenko, K.; Darrell, T. Adversarial discriminative domain adaptation. In Proceedings of the Computer Vision and Pattern Recognition (CVPR), Honolulu, HI, USA, 21–26 July 2017; Volume 1, p. 4.
18. Luo, Z.; Zou, Y.; Hoffman, J.; Fei-Fei, L.F. Label efficient learning of transferable representations acrosss domains and tasks. In Proceedings of the Advances in Neural Information Processing Systems, Long Beach, CA, USA, 4–9 December 2017; pp. 165–177.
19. Goodfellow, I.; Pouget-Abadie, J.; Mirza, M.; Xu, B.; Warde-Farley, D.; Ozair, S.; Courville, A.; Bengio, Y. Generative adversarial nets. In Proceedings of the Advances in Neural Information Processing Systems, Montreal, QC, Canada, 8–13 December 2014; pp. 2672–2680.
20. Isola, P.; Zhu, J.Y.; Zhou, T.; Efros, A.A. Image-to-Image Translation with Conditional Adversarial Networks. In Proceedings of the 2017 IEEE Conference on Computer Vision and Pattern Recognition (CVPR), Honolulu, HI, USA, 21–26 July 2017. [CrossRef]
21. Patel, V.M.; Gopalan, R.; Li, R.; Chellappa, R. Visual Domain Adaptation: A survey of recent advances. *IEEE Signal Process. Mag.* **2015**, *32*, 53–69. [CrossRef]
22. Saenko, K.; Kulis, B.; Fritz, M.; Darrell, T. Adapting visual category models to new domains. In Proceedings of the European Conference on Computer Vision, Heraklion, Crete, Greece, 5–11 September 2010; Springer: Berlin/Heidelberg, Germany, 2010; pp. 213–226.
23. Ganin, Y.; Ustinova, E.; Ajakan, H.; Germain, P.; Larochelle, H.; Laviolette, F.; Marchand, M.; Lempitsky, V. Domain-adversarial training of neural networks. *J. Mach. Learn. Res.* **2016**, *17*, 2096–2030.
24. Ganin, Y.; Lempitsky, V. Unsupervised domain adaptation by backpropagation. *arXiv* **2014**, arXiv:1409.7495.
25. Bousmalis, K.; Silberman, N.; Dohan, D.; Erhan, D.; Krishnan, D. Unsupervised pixel-level domain adaptation with generative adversarial networks. In Proceedings of the IEEE Conference on Computer Vision and Pattern Recognition, Honolulu, HI, USA, 21–26 July 2017; pp. 3722–3731.
26. Hoffman, J.; Tzeng, E.; Park, T.; Zhu, J.Y.; Isola, P.; Saenko, K.; Efros, A.A.; Darrell, T. Cycada: Cycle-consistent adversarial domain adaptation. *arXiv* **2017**, arXiv:1711.03213.
27. Ros, G.; Sellart, L.; Materzynska, J.; Vazquez, D.; Lopez, A.M. The synthia dataset: A large collection of synthetic images for semantic segmentation of urban scenes. In Proceedings of the IEEE conference on Computer Vision and Pattern Recognition, Las Vegas, NV, USA, 27–30 June 2016; pp. 3234–3243.
28. Vazquez, D.; Lopez, A.M.; Marin, J.; Ponsa, D.; Geronimo, D. Virtual and real world adaptation for pedestrian detection. *IEEE Trans. Pattern Anal. Mach. Intell.* **2014**, *36*, 797–809. [CrossRef] [PubMed]
29. Peng, X.; Saenko, K. Synthetic to real adaptation with generative correlation alignment networks. In Proceedings of the 2018 IEEE Winter Conference on Applications of Computer Vision (WACV), Lake Tahoe, NV, USA, 12–15 March 2018; pp. 1982–1991.

30. Shrivastava, A.; Pfister, T.; Tuzel, O.; Susskind, J.; Wang, W.; Webb, R. Learning from simulated and unsupervised images through adversarial training. In Proceedings of the IEEE Conference on Computer Vision and Pattern Recognition, Honolulu, HI, USA, 21–26 July 2017; pp. 2107–2116.
31. Shafaei, A.; Little, J.J.; Schmidt, M. Play and learn: Using video games to train computer vision models. *arXiv* **2016**, arXiv:1608.01745.
32. Hoffman, J.; Wang, D.; Yu, F.; Darrell, T. Fcns in the wild: Pixel-level adversarial and constraint-based adaptation. *arXiv* **2016**, arXiv:1612.02649.
33. Zhang, Y.; David, P.; Gong, B. Curriculum domain adaptation for semantic segmentation of urban scenes. In Proceedings of the IEEE International Conference on Computer Vision, Venice, Italy, 22–29 October 2017; pp. 2020–2030.
34. Sankaranarayanan, S.; Balaji, Y.; Jain, A.; Lim, S.N.; Chellappa, R. Unsupervised domain adaptation for semantic segmentation with gans. *arXiv* **2017**, arXiv:1711.06969.
35. Tsai, Y.H.; Hung, W.C.; Schulter, S.; Sohn, K.; Yang, M.H.; Chandraker, M. Learning to adapt structured output space for semantic segmentation. In Proceedings of the IEEE Conference on Computer Vision and Pattern Recognition, Salt Lake City, UT, USA, 18–22 June 2018; pp. 7472–7481.
36. Huang, H.; Huang, Q.; Krahenbuhl, P. Domain transfer through deep activation matching. In Proceedings of the European Conference on Computer Vision (ECCV), Munich, Germany, 8–14 September 2018; pp. 590–605.
37. Tasar, O.; Happy, S.L.; Tarabalka, Y.; Alliez, P. ColorMapGAN: Unsupervised Domain Adaptation for Semantic Segmentation Using Color Mapping Generative Adversarial Networks. *arXiv* **2019**, arXiv:1907.12859.
38. Fang, B.; Kou, R.; Pan, L.; Chen, P. Category-Sensitive Domain Adaptation for Land Cover Mapping in Aerial Scenes. *Remote. Sens.* **2019**, *11*, 2631. [CrossRef]
39. Gerke, M. *Use of the Stair Vision Library Within the ISPRS 2D Semantic Labeling Benchmark (Vaihingen)*; University of Twente: Enschede, The Nerthands, 2014.
40. Oliehoek, F.A.; Savani, R.; Gallego, J.; van der Pol, E.; Gross, R. Beyond Local Nash Equilibria for Adversarial Networks. *arXiv* **2018**, arXiv:1806.07268.
41. Goodfellow, I.J. NIPS 2016 Tutorial: Generative Adversarial Networks. *arXiv* **2016**, arXiv:1701.00160.
42. Liu, M.Y.; Breuel, T.; Kautz, J. Unsupervised Image-to-Image Translation Networks. In Proceedings of the Advances in Neural Information Processing Systems 30 (NIPS 2017), Long Beach, NV, USA, 4–9 December 2017.
43. Zhu, J.Y.; Zhang, R.; Pathak, D.; Darrell, T.; Efros, A.A.; Wang, O.; Shechtman, E. Toward Multimodal Image-to-Image Translation. In Proceedings of the Advances in Neural Information Processing Systems 30 (NIPS 2017), Long Beach, NV, USA, 4–9 December 2017.
44. Yi, Z.; Zhang, H.; Tan, P.; Gong, M. DualGAN: Unsupervised Dual Learning for Image-to-Image Translation. In Proceedings of the 2017 IEEE International Conference on Computer Vision (ICCV), Venice, Italy, 22–29 October 2017; pp. 2868–2876.
45. Zhu, J.Y.; Park, T.; Isola, P.; Efros, A.A. Unpaired Image-to-Image Translation Using Cycle-Consistent Adversarial Networks. In Proceedings of the 2017 IEEE International Conference on Computer Vision (ICCV), Venice, Italy, 22–29 October 2017. [CrossRef]
46. Xu, B.; Wang, N.; Chen, T.; Li, M. Empirical Evaluation of Rectified Activations in Convolutional Network. *arXiv* **2015**, arXiv:1505.00853.
47. Ioffe, S.; Szegedy, C. Batch Normalization: Accelerating Deep Network Training by Reducing Internal Covariate Shift. *arXiv* **2015**, arXiv:1502.03167.
48. Srivastava, N.; Hinton, G.E.; Krizhevsky, A.; Sutskever, I.; Salakhutdinov, R.R. Dropout: a simple way to prevent neural networks from overfitting. *J. Mach. Learn. Res.* **2014**, *15*, 1929–1958.
49. Mirza, M.; Osindero, S. Conditional Generative Adversarial Nets. *arXiv* **2014**, arXiv:1411.1784.
50. Yu, C.; Wang, J.; Peng, C.; Gao, C.; Yu, G.; Sang, N. BiSeNet: Bilateral Segmentation Network for Real-Time Semantic Segmentation. In Proceedings of the European Conference on Computer Vision, Lecture Notes in Computer Science, Munich, Germany, 8–14 September 2018; pp. 334–349. [CrossRef]
51. Real-Time Semantic Segmentation on Cityscapes. Available online: https://paperswithcode.com/sota/real-time-semantic-segmentation-cityscap (accessed on 28 March 2019).
52. Semantic Segmentation Suite. Available online: https://github.com/GeorgeSeif/Semantic-Segmentation-Suite (accessed on 28 March 2019).

53. Abadi, M.; Barham, P.; Chen, J.; Chen, Z.; Davis, A.; Dean, J.; Devin, M.; Ghemawat, S.; Irving, G.; Isard, M.; et al. TensorFlow: A system for large-scale machine learning. In Proceedings of the 12th USENIX Symposium on Operating Systems Design and Implementation (OSDI 16), Savannah, GA, USA, 2–4 November 2016; pp. 265–283.
54. He, K.; Zhang, X.; Ren, S.; Sun, J. Deep Residual Learning for Image Recognition. In Proceedings of the 2016 IEEE Conference on Computer Vision and Pattern Recognition (CVPR), Las Vegas, NV, USA, 26 June–1 July 2016. [CrossRef]
55. Kingma, D.P.; Ba, J. Adam: A Method for Stochastic Optimization. *arXiv* **2014**, arXiv:1412.6980.

© 2020 by the authors. Licensee MDPI, Basel, Switzerland. This article is an open access article distributed under the terms and conditions of the Creative Commons Attribution (CC BY) license (http://creativecommons.org/licenses/by/4.0/).

Article

Semantic Component Association within Object Classes Based on Convex Polyhedrons

Petra Đurović *,†, Ivan Vidović and Robert Cupec †

Faculty of Electrical Engineering, Computer Science and Information Technology Osijek, 31000 Osijek, Croatia; ivan.vidovic@ferit.hr (I.V.); robert.cupec@ferit.hr (R.C.)
* Correspondence: petra.durovic@ferit.hr; Tel.: +385-98-175-1176
† These authors contributed equally to this work.

Received: 12 February 2020; Accepted: 7 April 2020; Published: 11 April 2020

Featured Application: The application of the proposed work is in robotics. If a certain robot operation is defined for a part of a particular object, it can be transferred to other class instances by applying the proposed method for semantic association of the object components.

Abstract: Most objects are composed of semantically distinctive parts that are more or less geometrically distinctive as well. Points on the object relevant for a certain robot operation are usually determined by various physical properties of the object, such as its dimensions or weight distribution, and by the purpose of object parts. A robot operation defined for a particular part of a representative object can be transferred and adapted to other instances of the same object class by detecting the corresponding components. In this paper, a method for semantic association of the object's components within the object class is proposed. It is suitable for real-time robotic tasks and requires only a few previously annotated representative models. The proposed approach is based on the component association graph and a novel descriptor that describes the geometrical arrangement of the components. The method is experimentally evaluated on a challenging benchmark dataset.

Keywords: component association; semantic segmentation; part recognition

1. Introduction

One of the trends in robotics is to reduce the need for robot programing by allowing a robot to learn certain tasks from a human instructor. One approach to this problem is kinesthetic training of a robot, where a human manually guides a robot manipulator to perform certain action and then the robot applies the learned action to solve a practical task [1,2]. An advanced version of such training would be to define a robot action for a particular instance of an object class, referred to in this paper as a representative object, and apply an algorithm that would adapt this action to the other instances of the same class. In order to achieve this capability, the considered algorithm must associate the components of the representative object relevant for a particular task with the corresponding segments of the other instances of the same object class. The problem addressed in this paper is how to associate the components of different objects that have the same purpose. Since the target application field considered in this paper is robotics, we define components as regions of the object's surface that could potentially represent contact surfaces between the object and a robot tool when performing some task. For example, if the task is to carry a mug, it should be grasped by the handle. For a light bulb changing task, a robot should be able to recognize the light bulb. A usual approach for semantic segmentation of objects is to train the algorithm on manually annotated training and validation datasets and test it using a test dataset. Since the effort of manually annotating training data is time and energy consuming, the motivation for this paper was minimization of such labor. Without any prior knowledge about the purpose of a particular object class, the only cues that can be used to solve this problem are the similarity of the shapes of components and their spatial arrangement. The real-time execution of the algorithm is crucial for practical robotic application.

1.1. The Problem and the Contributions

The problem addressed in this paper is more precisely stated as follows. Let us consider a database of 3D models of objects belonging to the same class, represented by triangular meshes. There is a number of such datasets that are publicly available. An algorithm selects a small subset of representative objects from this dataset. On each representative object, a human expert annotates the target component specific for a certain task. According to these annotations, the algorithm identifies the semantically corresponding component of each of the remaining models in the database, referred to in this paper as query objects. The proposed approach enables easy and fast expansion of the existing model database with new query objects. In an ideal case, the annotation of a component on a single representative object should be sufficient for identification of all corresponding components in a given object class. However, certain object classes can comprehend objects whose components differ significantly in their shape, size, and position. Therefore, often more than one annotated representative object is required. The focus of our research is computational efficiency, required for practical applications in robotics. Our goal is to develop a method that identifies the target component on a newly perceived object and add this object to the existing database in a few seconds.

The approach presented in this paper is based on the detection of convex and concave surfaces, referred to in this paper as segments. Components are either represented by one or multiple segments. A selected component of the representative object is associated with the corresponding segments of the other instances of the same object class by constructing a component association graph (CAG). Graph nodes represent all segments of all models from the model database. The nodes are interconnected by edges. The weight of each edge represents a measure of the likelihood that these two segments belong to object components that have the same purpose. This measure is comprised of segment size, shape, position, and neighborhood similarity. The proposed similarity measures are based on the convex template instance (CTI) descriptor, proposed in [3], which describes object segments by approximating their shape with convex polyhedrons. A neighborhood similarity is defined by a novel descriptor, named the topological relation descriptor (TRED), which describes the topological relations between two segments in the model, which is also based on the CTI descriptor. Three methods for associating the selected component of the representative models with the segments of the other instances of the same object class using the CAG are proposed in this paper. The direct segment association method associates each segment according to its nearest neighbor in the CAG. The object-constrained association method associates the segments of the query object with each representative objects using a greedy search and computes the matching score between the query object and all representative object. The associations established between the query object and the representative object with the highest matching score are taken as the final result. The MST-based association method associates the query object segments with the representative object segments based on a minimum spanning tree (MST). As the result of any of these three methods, the selected component of the representative model defined for a certain task is associated with the corresponding segments of all models from the model database.

Accordingly, the following contributions of this paper are proposed.

1. A novel computationally efficient approach for establishing associations between components of an object of a given class, based on the component association graph.
2. A novel topologicalrelation descriptor (TRED), which describes the geometrical arrangement of components in a 3D object model.

1.2. Paper Overview

The paper is structured as follows. In Section 2, the related research is presented. Sections 3–6 describe the proposed methodology. Section 3 provides a formal problem definition, an overview of the proposed approach and an explanation of the CAG. In Section 4, an approach for measuring the semantic correspondence likelihood, assigned to the edges of the component association graph, is provided. In Section 5, three methods for the final association between the segments and a target component, based on the CAG, are proposed. Section 6 describes a method for selecting the representative objects. An experimental evaluation and a discussion of the results are given in Section 7. The paper is concluded in Section 8.

2. Related Research

Semantic segmentation in 3D was recently established as an important task in various applications: autonomous driving, human-machine interaction, object manipulation, manufacturing, geometric modeling, and reconstruction of 3D scenes, just to name a few. To facilitate the development of 3D shape understanding, the ShapeNet challenge [4] for semantic 3D shape segmentation on a large-scale 3D shape database was proposed. In this challenge, ShapeNet Parts, a subset of 16 classes from the ShapeNet database [5], was used, which is also used in the experimental analysis reported in this paper. The PointCNN method introduced in the ShapeNet challenge was later improved in [6]. It represents a generalization of the typical convolutional neural network (CNN) architecture for feature learning from point clouds. The segmentation accuracy of the PointCNN, experimentally evaluated on ShapeNet Parts dataset, outperformed 14 other methods in segmenting objects belonging to seven classes, while it achieved comparable accuracy in segmentation of objects from the other nine classes. The CNN architectures have been shown to be the most accurate and efficient approaches in a recent review of deep learning techniques applied to semantic segmentation, given in [7]. The best accuracy of object segmentation achieved by seven deep learning methods reported in [4] was between 63 and 96%, depending on the object class.

However, manual annotation of large datasets required for training neural networks is a time-consuming and delicate problem. Therefore, Yi et al. [8] proposed a novel active learning method capable of segmenting massive geometric datasets into accurate semantic regions that grants a good accuracy vs. efficiency trade-off. The ShapeNet Parts dataset was annotated using this approach. The goal of this approach (reducing human work required for the annotation of large datasets) is also the main motivation of the component association method proposed in this paper. The framework [8] achieved the annotation of large-scale datasets by cycling between manually annotating the regions, automatically propagating these annotations across the rest of the shapes, manually verifying both human and automatic annotations, and learning from the verification results to improve the automatic propagation algorithm. Although the approach proposed in [8] included the propagation of component labels from a small object set to a larger set, the focus of their research was on the whole iterative annotation process. A kind of continuation of the research [8] was given in [9]. The method also propagates labels from a small subset to a large dataset by global optimization analogous to the approach proposed in [10], which is based on conditional random fields. Our research, on the other hand, focuses on efficient detection of the target component on a query object given a small annotated set of representative objects, which could allow database expansion in real time.

An automatic approach to achieve semantic annotation of 3D models was proposed in [11]. The approach extracts concave and convex features as the cues for object decomposition into structural parts. By analyzing the position, shape, size, and configuration of the structural parts, the semantic category is assigned to each of them. The proposed methodology resembles the approach proposed by our paper, but it is applied in the semantic annotation of architectural buildings; therefore, the descriptors are adapted to that application, e.g., relative height, volume, dimension ratio, form mode, etc. The final assignment of the semantic label to a part in [11] was performed by the decision tree and the adapted support vector machine, while in this paper, three variants of assigning the label based on the CAG are proposed.

Analogous to the related methods discussed in this section, our method also performs segmentation of complete 3D object models. However, in order to use the algorithm proposed in this paper in a practical robotic application, a full 3D model of a given query object must be inferred from sensor data. The Pixel2Mesh deep neural network [12], AtlasNet [13], and occupancy network (ONet) [14] reconstruct full 3D object models from RGB images. AtlasNet and ONet can also receive 3D point clouds as the input. The generative shape proposal network (GSPN) [15] processes point clouds for the purpose of solving the semantic segmentation problem. This approach deals with cluttered real indoor scenes, partial point clouds, and point clouds of part-segmented 3D models, which represent real case scenarios in robotic applications.

3. Overview of the Proposed Approach

Let M be a set of 3D models of objects belonging to the same semantic class. This model set is referred to in this paper as a model database. It contains 3D models of objects represented by triangular meshes $P_k, k = 1, ..., n_M$. The considered algorithm should select a small set R of representative objects and present them to a human expert, which is asked to annotate a component relevant for a particular task on every object in this set. This annotation assumes the selection of a subset of mesh vertices of each mesh $P_r \in R$, referred to in this paper as points. The algorithm should then automatically label each mesh $P_k \in M \backslash R$, by assigning the label 1 to the points representing the target component and 0 to the remaining points.

3.1. Component Detection

In the approach proposed in this paper, object components are detected by segmenting the object's surface into convex and concave segments. This segmentation can be performed using the method proposed in [16]. These segments represent component proposals. The segmentation is performed by segmenting the model mesh into planar patches using the method applied in [17] and aggregating these patches according to the convexity criterion. In this paper, the term concave surfaces is used to denote inverted convex surfaces, i.e., convex surfaces with opposite local surface normals, as illustrated by Figure 1. Each segment of each model in a model database is assigned a unique ID representing a pair of indexes (i, k) where i denotes the segment index and k the model index. One semantic component can be represented by multiple convex or concave segments. For example, the mug handle shown in Figure 2 is represented by one convex and one concave surface.

Since the segmentation of certain shapes can be ambiguous, some additional segments obtained by merging the original segments are created, in order to cover a variety of possible segmentation variants. The algorithm applied for this segment merging is described in Appendix A.

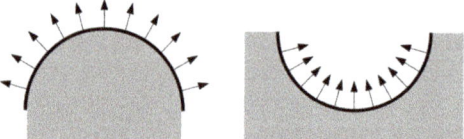

Figure 1. An example of a convex (**left**) and a concave (**right**) surface with surface normals denoted by arrows.

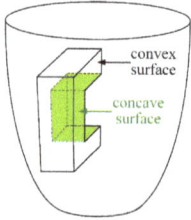

Figure 2. Representation of a mug handle by one convex and one concave surface.

3.2. Component Association Graph

The method for establishing associations between segments of the objects of the same class, proposed in this paper, is based on the component association graph (CAG). Nodes of the CAG represent segments of all models from the database, connected by edges with the assigned semantic correspondence likelihood measure (SCLM). Each segment of the query object is, thus, connected with n nodes with the greatest SCLM values, referred to in this paper as the nearest neighbors. The number of nodes is limited in order to reduce the computational complexity, which is especially important in the case of large databases. The SCLM consists of the segment shape, size, position, and neighborhood similarity measure. Three methods for establishing the final associations between segments and the target component based on the constructed CAG are proposed in Section 5. As a result of this association process, each segment in the model database is assigned a label with a value of one, if this segment is associated with the target component, or a value of zero otherwise.

4. Semantic Correspondence Likelihood

The likelihood that a query object segment C_j^Q and the kth model segment C_{ik}^M represent semantically corresponding components is assessed by the SCLM computed by:

$$y_{ijk} = y_{ijk}^C + w_N y_{ijk}^N, \qquad (1)$$

where y_{ijk}^C represents segment shape, size, and position similarity, y_{ijk}^N represents the segment neighborhood similarity, and w_N is a weighting factor. The computation of the shape, size, position, and neighborhood similarity is described in the following subsections.

4.1. CTI Descriptor

The computation of the SCLM is based on the CTI descriptor proposed in [3]. In this subsection, a brief description of this descriptor is provided. Let A be a set of different unit vectors $a_m \in \mathbb{R}^3, m = 1, ..., n_a$ representing standardized normals. Furthermore, let us consider a set of all convex polyhedrons such that each face of the polyhedron is perpendicular to one of the vectors $a_m \in A$. The set A is referred to as a convex template, and each convex polyhedron belonging to the considered polyhedron set is referred to as a convex template instance (CTI). CTI is uniquely defined by a convex template A and a vector $d = [d_1, d_2, ..., d_{n_a}]$, where d_m represents the distance between the mth polyhedron face and the origin of the object RF. This vector represents the CTI descriptor. The approach proposed in this paper requires that for each unit vector $a_m \in A$, there exist its opposite vector $a_{\overline{m}} \in A$. The CTI descriptor was originally designed for fruit recognition [3]. Later, it was applied for the alignment of similar shapes with the purpose of object classification on depth images [18]. In [16], the CTI descriptor was applied for solving the shape instance detection problem. A CTI descriptor is computed for each object segment. Four examples of objects represented by CTIs are shown in Figure 3.

4.2. Segment Shape, Size, and Position Similarity

The shape, size, and position similarity of two segments is measured by comparing their CTI descriptors. In order to reduce computational complexity, the CTI descriptors, d, are projected onto a lower dimensional latent space, as proposed in [19]. Thereby, descriptors q of $n_q < n_d$ elements are obtained by:

$$q = O^T d, \qquad (2)$$

where O represents an orthonormal basis defining a latent space computed by performing the principal component analysis (PCA) of the CTI descriptors d extracted from the segments of a training set. Another reason for computing the latent vectors is to decouple the shape from the position information. The first three elements of q, denoted in this paper by q^t, represent the position of the segment in the object RF, while the other 21 elements, denoted in this paper by q^s, describe its shape and size. Let us consider a model database M of n_M models. The segments of the kth model are represented by latent vectors $q_{ik}^M, i = 1, ..., n_{M,k}$. Analogously, segments of the query object are represented by the latent vector $q_j^Q, j = 1, ..., n_Q$.

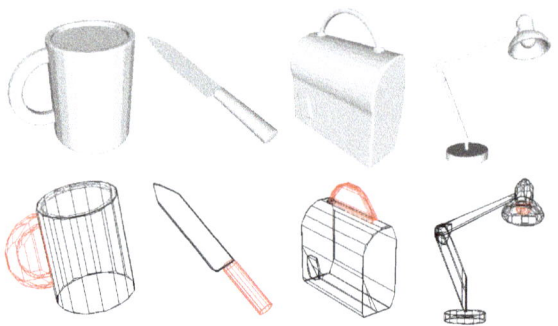

Figure 3. Representation of objects by convex template instances (CTIs).

The similarity between two segments taking into account their position (t), shape (s), and size (a) can be measured by Gaussian function:

$$y_{ijk}^C = \exp\left(-\frac{1}{2}\left(\frac{e_{ijk}^t}{\sigma_t^2} + \frac{e_{ijk}^s}{\sigma_s^2} + \frac{e_{ijk}^a}{\sigma_a^2}\right)\right), \quad (3)$$

where σ_t, σ_s, and σ_a represent parameters that define the contribution of the difference between the segment position, shape and size to the total similarity measure, respectively. The values of the algorithm parameters used in the experiments reported in Section 7 are given in that section. Translation, shape, and scale differences are computed by the following three equations.

$$e_{ijk}^t = \frac{||q_j^{Q,t} - q_{ik}^{M,t}||^2}{||q_j^{Q,s}||\,||q_{ik}^{M,s}||}, \quad (4)$$

$$e_{ijk}^s = 1 - \left(\left(\frac{q_j^{Q,s}}{||q_j^{Q,s}||}\right)^T \left(\frac{q_{ik}^{M,s}}{||q_{ik}^{M,s}||}\right)\right)^2, \quad (5)$$

$$e_{ijk}^a = \frac{\left(||q_j^{Q,s}|| - ||q_{ik}^{M,s}||\right)^2}{||q_j^{Q,s}||\,||q_{ik}^{M,s}||}. \quad (6)$$

The norm of vector q^s represents a measure of the segment size. In (4), the position difference is normalized by the size of the segments, therefore allowing bigger segments to have greater distance in order to achieve the same similarity measure values. Equation (5) contains a scalar product of two unit vectors. The greater the value of the scalar product, the more similar the shapes are. Equation (6) represents a segment size difference measure.

4.3. Neighborhood Similarity

Let us assume two segments C_i and C_j, described by CTI descriptors d_i and d_j, belonging to the same model. In order to describe the geometrical arrangement of segments in the model, we introduce the TRED that describes topological relations between two segments. The TRED represents a tuple $T(C_i, C_j) = (\mu_{ij}, v_{ij}, \sigma_{ij}^v)$. The relation type coefficient, μ_{ij}, is computed by:

$$\mu_{ij} = \min_{m=1,\dots,n_d} \rho_{ijm}, \qquad (7)$$

where:

$$\rho_{ijm} = \frac{d_{im} + d_{j\overline{m}}}{d_{im} + d_{i\overline{m}}}. \qquad (8)$$

In Equation (8), m and $\overline{m}$ represent the indexes of two opposite unit vectors $a_m, a_{\overline{m}} \in A$. Five types of topological relation between two segments C_i and C_j, defined by μ_{ij}, are considered:

Type 1: C_j contains C_i:

$$\mu_{ij} \geq 1 \wedge \mu_{ij} > \mu_{ji}$$

Type 2: C_i and C_j are identical:

$$\mu_{ij} = \mu_{ji} = 1$$

Type 3: C_i and C_j intersect:

$$0 < \mu_{ij} < 1 \wedge 0 < \mu_{ji} < 1$$

Type 4: C_i touches C_j:

$$\mu_{ij} = 0$$

Type 5: C_i and C_j are disjoint:

$$\mu_{ij} < 0.$$

The relation type between two segments is denoted in this paper by $type(T(C_i, C_j))$. Three of these relations are illustrated by Figure 4: a segment C_i is inside a segment C_j (left); two segments C_i and C_j intersect (middle); and a segment C_i touches a segment C_j (right).

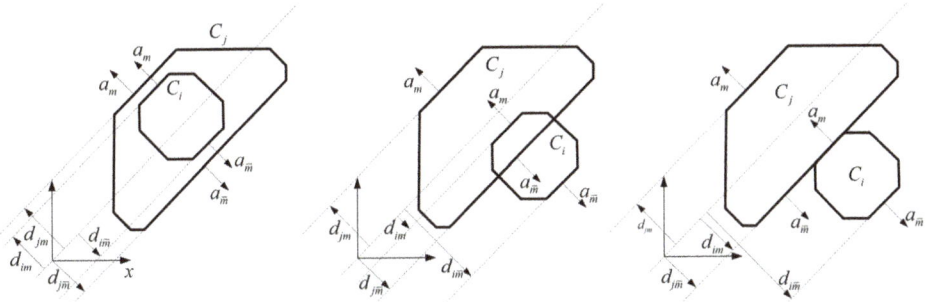

Figure 4. Topological relation descriptor (TRED). The CTI descriptor elements d_{xm} that are denoted with arrows directed opposite the corresponding vector a_m have negative values.

If two segments are touching (Type 4), v_{ij} and σ_{ij}^v are computed by:

$$v_{ij} = \frac{\sum\limits_{m=1}^{n_d} w_{ijm} a_m}{\left\| \sum\limits_{m=1}^{n_d} w_{ijm} a_m \right\|}, \qquad (9)$$

and:

$$\sigma_{ij}^v = \sqrt{\frac{\sum_{m=1}^{n_d} w_{ijk} \arccos^2 v_{ij}^T a_m}{\sum_{m=1}^{n_d} w_{ijk}} + \sigma_{v0}^2}, \quad (10)$$

where:

$$w_{ijm} = \exp\left(-\frac{1}{2}\left(\frac{\mu_{ij} - \rho_{ijm}}{\sigma_\rho}\right)^2\right). \quad (11)$$

Vector v_{ij} is approximately orthogonal to the plane that separates C_i and C_j, and it is directed from C_i to C_j. Its purpose is to describe the touching direction. This vector is computed as a weighted average of all vectors a_m, where vectors corresponding to smaller values ρ_{ijm} have greater weights. Value σ_{ij}^v describes the uncertainty of vector v_{ij}, where σ_{v0} is a constant that models the uncertainty due to the measurement noise. TREDs are computed for every pair of model segments. A segment C_j is a neighbor of a segment C_i if $type(T(C_i, C_j))$ is 1 or 4. In order to make the method robust to noise, the definitions of the topological relation types are relaxed by introducing a suitable tolerance. As a consequence of this relaxation, the topological relation of Type 2 includes not only identical segments, but also segments that occupy approximately the same space.

Furthermore, if two segments are of different convexity types (convex or concave), then they are considered to be neighbors only if their topological relation is of Type 1 or Type 2.

The neighbors of each segment are grouped into clusters according to the associated TRED descriptors. These clusters are denoted in this paper by Γ, and the cluster set assigned to a segment C_p^X is denoted by N_p^X, where X stands for the query (Q) or representative (M) model. Two neighbors of C_i are grouped into a cluster only if their topological relation to C_i is of the same type. In the case of the Type 4 relation, two neighbors C_j and C_k can be grouped in the same cluster if their vectors v_{ij} and v_{ik} are similar. Vectors v_{ij} and v_{ik} are considered to be similar if the angle between them is $\leq 60°$. In the case of the Type 1 relation, all neighbors C_j such that $\mu_{ij} > \mu_{ji}$ are grouped in one cluster and all neighbors C_j such that $\mu_{ij} < \mu_{ji}$ are grouped in another cluster. All neighbors of different convexity types than C_i whose topological relation is of Type 2 are grouped into one cluster.

Similarity between the neighborhood of a segment C_j^Q of a query object and a segment C_{ik}^M of the kth model is measured by matching their neighbor clusters. Let C_m^Q and C_{lk}^M be neighbors of C_j^Q and C_{ik}^M, respectively. The similarity between these two segments is measured by:

$$y_{lmk}^n = \varepsilon_{lmk} \exp\left(-\frac{1}{2}\left(\frac{e_{lmk}^s}{\sigma_s^2} + \frac{e_{lmk}^a}{\sigma_a^2}\right)\right), \quad (12)$$

where ε_{lmk} is a binary variable, which has a value of one only if:

$$type(T(C_j^Q, C_m^Q)) = type(T(C_{ik}^M, C_{lk}^M)),$$

or a value of zero otherwise. Furthermore, two neighbors C_m^Q and C_{lk}^M of Type 4 are matched only if vectors v_{jm} and v_{ilk} of the associated TREDs have sufficiently similar directions taking into consideration their uncertainties described by σ_{jm}^v and σ_{ilk}^v. Otherwise, $\varepsilon_{lmk} = 0$.

The similarity of two neighbor clusters $\Gamma_{rj}^Q \in N_j^Q$ and $\Gamma_{pik}^M \in N_{ik}^M$, where p and r represent cluster indexes, is measured by the similarity of the shape and size of the two most similar segments belonging to these two clusters. The similarity measure of two clusters is computed by:

$$y^{\Gamma}(\Gamma_{rj}^Q, \Gamma_{pik}^M) = \max_{\substack{C_m^Q \in \Gamma_{rj}^Q \\ C_{lk}^M \in \Gamma_{pik}^M}} y_{lmk}^n.$$

The neighborhood similarity measure is computed by applying the following procedure. First, an initially empty set B_{ijk}^N is created. Then, the pair of clusters with the highest value y^{Γ} is identified and stored in B_{ijk}^N. This step is repeated for all remaining clusters until no more cluster pairs can be formed. Finally, the resulting set B_{ijk}^N is used to compute the neighborhood similarity measure according to the following formula:

$$y_{ijk}^N = \sum_{(\Gamma,\Gamma') \in B_{ijk}^N} y^\Gamma(\Gamma, \Gamma').$$

5. Associating Segments with the Target Component

There are three proposed variants of assigning the target component to a query object segment based on the constructed CAG. Note that the construction of CAG is independent of the applied variant.

Direct Segment Association

The query object segment is assigned the ID of the nearest neighbor in the CAG representing a segment of a representative object.

Object-Constrained Association

The object-constrained association approach compares the query model with each representative model. Each segment of the query model is associated with every component of the representative model by a greedy search. The association with the greatest SCLM is detected, and the query segment is assigned the ID of the associated representative model component. That query segment is omitted from the further search, and the search continues until all query segments are associated with corresponding representative object segments. An object matching score between the query and the representative model is computed as the sum of SCLMs between the associated query model segments and representative model segments that represent the target component. Query model segments are finally assigned the IDs of the associated segments of the representative model with the greatest object matching score.

MST-Based Association

A root node is added to the CAG, which does not represent any segment. Then, a minimum spanning tree (MST) is constructed with the constraint that all segments of the representative objects are directly connected to the root node. Hence, each representative object segment of all representative models spreads one subtree. Segments of the query object are assigned the ID of the representative model segment in the same subtree.

At the end of any of the three proposed procedures, each query object segment is assigned the ID of a representative object segment. Finally, the query object segments inherit the labels of the associated representative object segments. The first and the third variant associate each query object segment with the ID of a corresponding representative object segment, independent of the labels of the representative object segments. Hence, these two methods can be performed before the target component is annotated on the representative objects.

6. Selection of Representative Objects

Let us consider a model database M representing a set of models M_k of objects belonging to the same object class. Each model M_k is represented by a 3D mesh. As previously explained, every mesh is segmented into convex and concave segments, and each of these segments is represented by a CTI descriptor. In order to facilitate the explanations of the proposed approach, a model M_k will be represented by a sequence of point sets $M_k = (C_{1k}, \ldots, C_{n_M,k}, P_k)$, where C_{ik} is the ith model segment and P_k is a set of 3D points lying on the surface of the kth model. For the computational efficiency, these points can be obtained by sub-sampling the mesh vertices. The similarity of the kth and the lth

object is assessed by measuring the distances between the points of the kth model to the CTIs of the lth model and opposite. The object similarity measure is computed by:

$$y_{kl}^O = y_{lk}^O = z_{kl}^O \cdot z_{lk}^O, \tag{13}$$

where:

$$z_{kl}^O = \frac{1}{|P_k|} \sum_{p \in P_k} \exp\left(-\frac{1}{2} \frac{\delta_{pM}^2(p, M_l)}{\sigma_p^2}\right) \tag{14}$$

$$\delta_{pM}(p, M_l) = \min_{C_{jl} \in M_l} (\delta_{pC}(p, C_{jl})) \tag{15}$$

$$\delta_{pC}(p, C_{jl}) = |\max_{m=1,\dots,n_d} (a_m^T p - d_{jlm})| \tag{16}$$

Parameter σ_p in (14) is an experimentally determined constant. If C_{jl} is a concave segment, then the *min* operation is used in (16) instead of *max*. Note that $y_{kl}^O \in [0,1]$.

Set R of the representative objects is selected by a greedy procedure that maximizes the similarity between R and all objects $M_k \in M$. The similarity between an object model $M_k \in M$ and set R is defined as the similarity between this model and the most similar model $M_r \in R$. This similarity can be measured by a value y_k^R computed by:

$$y_k^R = \max_{M_r \in R} y_{kr}^O.$$

The similarity between M and R can be measured by value y^{MR} representing the total sum of values y_k^R, i.e.,

$$y^{MR} = \sum_{k=1}^{|M|} y_k^R.$$

In the approach proposed in this paper, the set of representative objects R is selected by an iterative procedure, where in each iteration, a model that maximizes y_{MR} is selected until a predefined number of representative objects n_R is selected.

7. Experimental Evaluation

The proposed approach was experimentally evaluated using the ShapeNet 3D model database [5]. A part of this dataset is dedicated to testing methods that segment 3D models into semantic parts. This dataset consists of 16 subsets, each representing one object class. Each of these subsets is subdivided into training, validation, and test subsets, where the training subsets are significantly larger than the validation and test subsets. The dataset was originally designed to be used in the following way: the evaluated method should be trained using a manually annotated training dataset and the associated validation dataset and tested using the corresponding test dataset. However, in this paper, we investigated a different paradigm. We wanted to test the ability of the proposed approach to (i) segment unannotated 3D models of a particular object class into segments and establish associations between these segments according to the similarity of their shape, size, and geometric arrangement and (ii) to use the established associations to identify a user selected component in all models given a small set of manually annotated representatives. In order to automate the experiments that require a user selection, instead of manually annotating the representative set in each experiment, we used the ground truth annotations available for every model in the dataset.

Since the target application of the proposed approach was facilitating robotic operations, we selected six object classes from the considered dataset, which could be associated with an exactly defined robot task: mugs, knives, bags, lamps, caps, and laptops. For the mugs, knives, and bags, the target component was the handle for the robot grasping task. For the lamps, the target component was the light bulb for the replacement task. The target component of the cap was the peak for the dressing task, and the target component of the laptop was the screen for the opening and closing task.

In the case of all classes except the lamps, we used the original annotations of the ShapeNet dataset, where the object handle, peak, and screen were annotated as one of the semantic categories. In the case of lamps, we selected a subset of 210 lamps with clearly distinguishable light bulbs and annotated the light bulbs manually.

The following experimental procedure was applied to each of the six considered object classes, with the parameter values given in Table 1.

1. Every model was segmented into convex and concave segments using the approach proposed in [16].
2. Segment merging was performed by the procedure described in Appendix A.
3. Every segment was represented by a CTI descriptor. Analogously to [3,19], we used a convex template consisting of $n_d = 66$ unit vectors uniformly distributed on the unit sphere.
4. The TRED descriptor, described in Section 4, was computed for every segment pair of every model.
5. The neighbor clusters were identified and assigned to every segment.
6. The CTI descriptor of every segment was projected onto the corresponding latent space by (2).
7. The SCLM proposed in Section 4 was computed between each of two segments of different models in the model set.
8. A CAG was created. The nodes of this graph represented all segments of all models in the model set. Each node was connected to at most 100 most similar segments, according to the SCLM computed in Step 7.
9. A small number of representative objects was selected by the method proposed in Section 6. This number was ≤10% of all models in the considered model set.
10. A target component was annotated on the representative objects. In a practical application, this step would be performed manually by a human expert. In the reported experiments, the ground truth annotations, which were available for all models, were assigned to the representative object models.
11. All segments of all remaining models were automatically annotated using the CAG created in Step 8 and the three methods proposed in Section 5: direct segment association (DSA), object-constrained association (OCA), and MST-based association (MST).
12. The result obtained for every model was compared to the ground truth using the intersection over union (IoU) performance index.

Steps 1–8, the computation of object similarity in Step 9, and the creation of MST in Step 11 were implemented in C++. Step 12 and the rest of the Steps 9 and 11 were implemented in MATLAB.

Table 1. Parameter values.

n_d	n_q	σ_t	σ_s	σ_a	σ_{v0}	σ_p	τ_{tr}	τ_v
66	24	2.4	0.132	0.707	0.1	0.025	0.333	2

The computation of the latent vector, performed in Step 6, required an orthonormal basis defining a latent space. This orthonormal basis was represented by an $n_d \times n_q$ matrix O. This matrix was computed by the training procedure proposed in [19], where simpler CTI descriptors were used instead of the more complex VolumeNet descriptors, as proposed in [20]. The training was performed using the 3DNet dataset [21]. A total of 351 3D meshes of objects belonging to 10 classes were segmented into convex and concave surfaces, and a CTI descriptor was computed for each surface. Each CTI descriptor represented a point in an n_d-dimensional space. The obtained descriptors were collected in two subsets, one representing convex segments and the other concave segments. Matrix O

was computed for each descriptor set using the method based on the principal component analysis (PCA) proposed in [19]. For both convex and concave segments, we generated a latent space of $n_q = 24$ dimensions. This number of dimensions was chosen as the minimum number of the first principal components, such that the variances of the remaining principal components were $\leq 10^{-4}$ for both the convex and concave segment set.

The ground truth component annotation was available in the ShapeNet dataset for every model as a set of sample points with assigned labels. Since the proposed approach computed segment associations, the segments had to be associated with the labeled points. Each point was associated with the closest segment. Furthermore, the distance between every point and the CTI of every segment were computed, and every point was associated with a segment if the distance to the CTI of this segment was ≤ 0.001. This distance was computed by (16). Note that this was not the Euclidean distance, but it was a good approximation, which could be computed very efficiently. The same distance threshold could be used for all objects, since all models in the ShapeNet dataset were scaled to approximately the same size. This point-segment association allowed each point to be associated with more than one segment.

In Step 10 of the experimental procedure, a segment of a representative object was labeled as being part of the target component if more than half of the points it was associated with were annotated. Step 11 annotated the segments of all models from a considered model set, except the representative objects, which were already annotated in Step 10. In order to compute IoU in Step 12, annotations must be transferred from segments to sample points. A point was labeled as belonging to the target component if it was associated with at least one annotated segment.

7.1. Results

A few sample results are shown in Figure 5. True positives, i.e., points that were correctly associated with the target component, are depicted by the green color, while false positives, i.e., points that were falsely associated with the target component, are depicted in red color. The false negatives, representing the target component points, which were not associated with the target component by the proposed algorithm, are depicted in blue color.

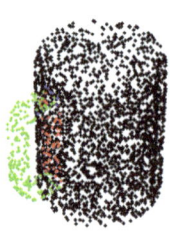  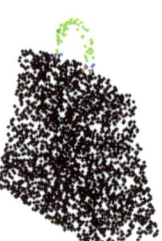 

Figure 5. Sample results.

The quantitative results obtained by the described experimental procedure are presented in Table 2. Each of the three component association methods were tested with and without using the neighborhood similarity, i.e., with weight w_N equal to one and zero, respectively. The number of representative objects was the greatest integer $\leq 10\%$ of all models in the model database of each object class. The presented IoU values represented the average over all objects of a particular class excluding the representative objects.

Table 2. Accuracy of the automatic component annotation. DSA, direct segment association; OCA, object-constrained association; MST, minimum spanning tree.

	Class		Mug	Knife	Bag	Lamp	Cap	Laptop
	total No. of objects		184	391	68	210	53	451
	No. of representatives		18	39	6	21	5	45
IoU	DSA	$w_N = 0$	78.50	75.40	44.10	43.10	82.50	75.30
		$w_N = 1$	77.40	73.30	50.30	57.20	84.80	77.20
	OCA	$w_N = 0$	73.90	58.00	48.00	53.60	82.40	58.90
		$w_N = 1$	75.10	62.40	43.60	65.40	82.40	59.30
	MST	$w_N = 0$	77.90	75.90	42.60	36.80	82.70	76.70
		$w_N = 1$	76.30	73.20	60.90	53.00	75.90	76.10

In order to compare the proposed method with the state-of-the-art approaches, the following experiment was executed. For each of the five classes: mugs, knives, bags, caps, and laptops, Steps 1–10 were performed on the training subset, which included the selection and annotation of representatives. Note that only a small representative subset of the training dataset was annotated. The number of annotated representatives was the greatest integer ≤10% of all models in the training subset of each object class. Then, the database was extended (Steps 1–8) by the test dataset; the labels were assigned to the models from this subset; and the accuracy analysis was performed (Steps 11–12). The results of the accuracy analysis performed on the test subset measured by per-class IoU are given in Table 3. At the bottom of the table, the accuracies achieved by the three methods (DSA, OCA, and MST) proposed in this paper are given. Since this paper investigated component association with a small set of representatives, it was not expected to outperform the existing methods extensively trained using large annotated datasets.

Table 3. Accuracy (%) comparison with the state-of-the-art methods.

Method/Class		Mug	Knife	Bag	Cap	Laptop
SyncSpecCNN [22]		92.73	86.10	81.72	81.94	95.61
Pd-Network [23]		94.00	87.25	82.42	87.04	95.44
SSCN [24]		95.23	89.10	82.99	83.97	95.78
SpiderCNN [25]		93.50	87.30	81.00	87.20	95.80
SO-Net [26]		94.20	83.90	77.80	88.00	94.80
PCNN [27]		94.80	86.00	80.10	85.50	95.70
KCNet [28]		94.40	87.20	81.50	86.40	95.50
Kd-Net [23]		86.70	87.20	74.60	74.30	94.90
3DmFV-Net [29]		94.00	85.70	84.30	86.00	95.20
RSNet [30]		92.60	87.00	86.40	84.10	95.40
DGCNN [31]		93.30	87.30	83.70	84.40	96.00
PointNet [32]		93.00	85.90	78.70	82.50	95.30
PointNet++ [33]		94.10	85.90	79.00	87.70	95.30
SGPN [34]		93.80	83.00	78.60	78.80	95.80
PointCNN [6]		95.28	88.44	86.47	86.04	96.11
DSA	$w_n = 0$	82.14	75.53	39.67	69.59	75.54
	$w_n = 1$	76.65	73.42	41.66	74.76	74.53
OCA	$w_n = 0$	75.74	56.35	36.24	81.45	59.67
	$w_n = 1$	76.47	59.18	40.34	80.28	59.95
MST	$w_n = 0$	77.68	75.59	34.92	75.57	76.74
	$w_n = 1$	75.80	64.78	44.01	72.32	65.91

Let us assume a database with the total number of objects as stated in Table 4. The running times of the automatic component annotation with the included neighborhood similarity ($w_N = 1$) in the SCLM are reported in Table 4. The average execution time of every step per object is presented. Furthermore, the total execution time of all steps required for expanding the model database with a new query object is provided. It was assumed that before the expansion, the database contained the total number of objects given in Table 4 minus one. The average running time per model is reported for Steps 1–8, implemented in C++, and for Steps 11 and 12, implemented in MATLAB. In the case of the MST method, the execution time of Step 11 included computation of MST implemented in C++. The representative selection time, Step 9, was required only when the representative models were selected. In the case where the database was extended without annotating a new representative object, this step was not performed. Thus, the execution times of this step are presented separately from the other steps. Furthermore, Step 10, which was performed by a human expert, is not considered in Table 4. The experiments were executed on a PC with an Intel Core i7-4790 3.60GHz processor and 16 GB of installed RAM memory running Windows 10 64-bit OS.

Table 4. Running time of the automatic component annotation.

Class					Mug	Knife	Bag	Lamp	Cap	Laptop
total no. of objects					184	391	68	210	53	451
no. of representatives					18	39	6	21	5	45
average step execution time per object [ms]	C++	steps 1–6			825	1112	1484	917	702	288
		steps 7–8			19	70.6	36.2	62.4	3.8	14.6
	MATLAB	steps 11–12	DSA		4.7	10.6	5.6	8.2	2.1	3.5
			OCA		109.7	667.5	11.9	199.1	4.6	383.4
			MST		52.2	106.7	14.9	45.4	5.3	72.5
total time per object [ms]				min	849	1193	1526	988	708	306
				max	954	1850	1535	1179	711	686
representative selection time [ms]	C++	step 9			59,847	198,611	12,249	82,930	4873	314,175
	MATLAB	step 9			10	26	2	4	1	35

In order to investigate how the component association accuracy depended on the number of representative objects, we performed a series of experiments, where we varied the number of representatives from one to the value shown in Table 2. The results of this analysis are shown in Figure 6. The experiments were performed for all three component association methods with $w_N = 1$.

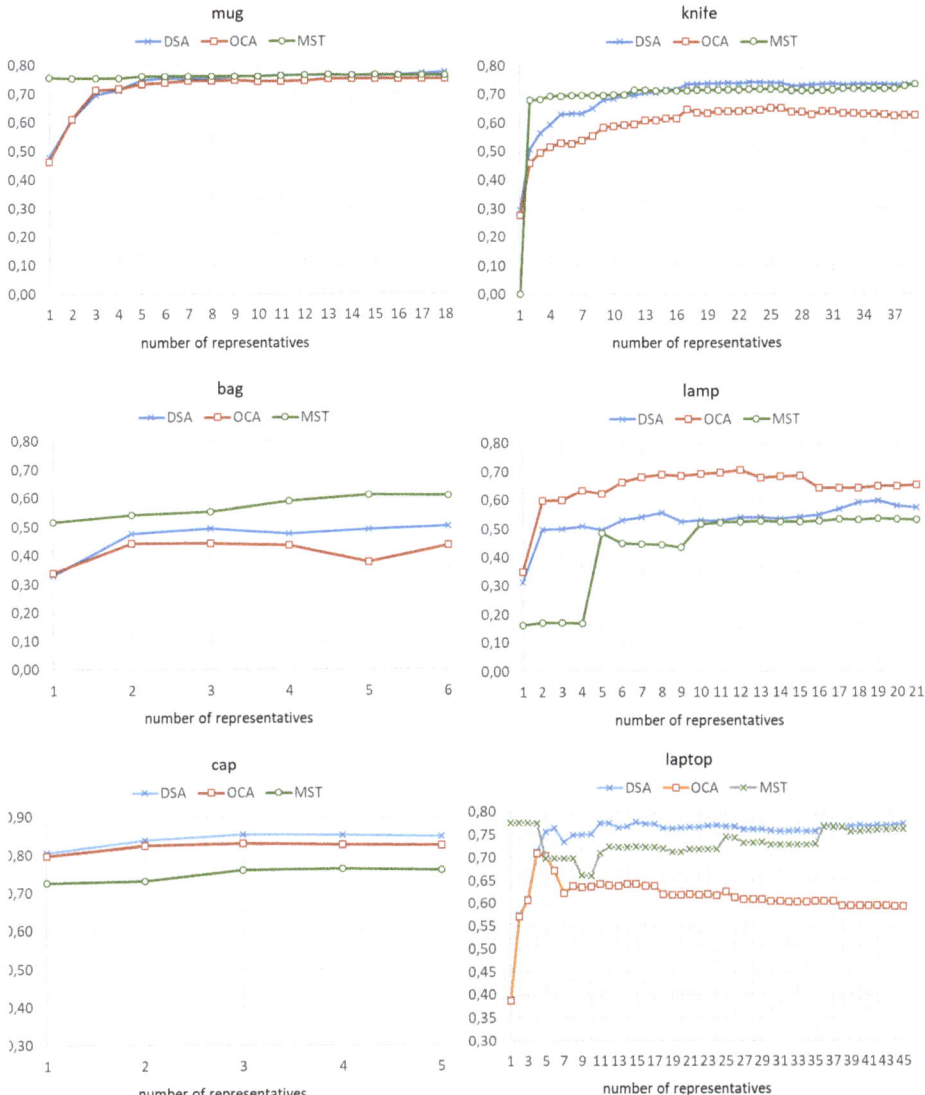

Figure 6. Component association accuracy depending on the number of representatives and the applied method measured by IoU.

7.2. Discussion

The standard pipeline for semantic object segmentation with standard machine learning or deep neural networks consists of training, validation, and test parts. The training is usually performed on a large subset containing many annotated objects. The time spent for the annotation and training is rarely reported in the research, but it surely requires much time and energy. This experimental analysis investigated how accurate semantic association can be achieved without extensive training and annotation. The reported results proved that only a few annotated representative models were required to achieve the reported accuracy. In the case of simple objects, such as knives, caps, mugs, and laptops, only one or two annotated models were required to achieve an IoU of approximately 0.7.

From the presented experimental analysis, it could be concluded that including neighborhood similarity in the SCLM improved the results significantly in the case of bags and lamps, which had more complex shapes than mugs, knives, caps, and laptops. In the case of mugs, knives, caps, and laptops, which had simple shapes, the information about the topological relations between components did not enhance the ability of the algorithm to identify the target component.

For all classes except the lamps, the MST component association method would be the method of choice, because it provided a relatively high IoU for a small number of representatives, which did not grow significantly with the number of representatives. In the case of mugs, caps, and laptops, a single representative was sufficient to identify the target component in the remaining objects with IoU > 0.70. In the case of knives, the wrong selection of the first representative by the proposed method resulted in a very low IoU. However, with the second representative, an IoU close to 0.7 was reached.

The MST method was expected to provide better results than the DSA in the case where the model database represented a sufficiently dense sampling of the considered object class. We considered a model database to be dense if for any two objects, a sequence of objects could be found where two consecutive objects had sufficiently similar shapes to allow unambiguous segment association. An example of such a sequence is shown in Figure 7. A bag handle of a reference object, shown in the top left image of Figure 7, was correctly associated with the segment represented by green points in the bottom right image of Figure 7. In an ideal case, the processing of such a model database by the MST method would allow correct component association using a single annotated representative object. However, in order for a dataset to be dense and cover large shape variations, it has to be very large. In order to evaluate how the size of the object dataset affected the accuracy of the MST-based component association, the procedure explained in Steps 1–12 was performed for all six classes with different numbers of models in the database, varying from 25% to 100% of the total No. of objects given in Table 2 and with the included neighborhood similarity, $w_N = 1$. The results are given in Table 5.

Figure 7. An example of the dense database with similar consecutive objects.

Table 5. Accuracy of the component annotation by MST with different numbers of objects in the database.

Percent of Database/Class	Mug	Knife	Bag	Lamp	Cap	Laptop
25%	0.721	0.755	0.400	0.080	0.706	0.762
50%	0.747	0.726	0.419	0.177	0.732	0.694
75%	0.752	0.749	0.604	0.400	0.793	0.739
100%	0.763	0.732	0.609	0.530	0.759	0.761

Although the proposed method provided interesting results for a small numbers of representative objects, it did not achieve very high accuracy even when a high percentage of annotated representative objects was used. One of the reasons was that the proposed approach relied on a segmentation method that provided segments that often crossed the boundary of the target component. This was noticeable in the case of mug and bag handles, where the concave segment representing the inner side of a handle extended to the body of the object, resulting in false positives, as illustrated by the example shown in the leftmost image of Figure 5, where the red points were falsely assigned to the handle.

The goal of the proposed approach was to reduce the annotation effort and allow an easy expansion of the model dataset by new models in real time. The execution times provided in the Table 4 indicated that expanding the model database with a new model, recomputation of the CAG, and establishing the component-segment correspondences required between approximately 0.3 and 1.9 seconds, depending on the complexity of the object structure, the chosen method, and the number of objects already in the model dataset. Thus, the proposed method is usable for real time robotic applications. The active learning framework for annotating massive geometric datasets [8] reported the average time required for preprocessing and annotation per shape of approximately one minute for a database of 400 objects. This time was expected to scale linearly with the size of the database. Furthermore, the analysis of a new shape, in a method for converting geometric shapes into hierarchically segmented labeled parts, proposed by [9], typically took about 25 seconds. Comparing with these two methods, which also propagated labels from a small subset to a large dataset, the method proposed in this paper is significantly faster.

8. Conclusions

The aim of the paper was to investigate how accurate semantic association can be achieved without extensive training and annotation of a large amount of data and if such an approach could be time effective for real-time applications. The target application of the proposed approach was in robotics, where it could be used in combination with an object detection and 3D reconstruction module. The task of this module would be to detect objects of a particular class in an RGB-D image or a LiDAR scan and to reconstruct their full 3D model. The proposed approach could then be used to associate the components of the reconstructed object, with the corresponding components of representative object models. In order to be applicable in real-world scenarios, the robot perception system must be able to cope with cluttered scenes, where the target objects are partially visible and appear in various poses. Augmentation of the proposed approach with a method that would detect objects of a given class in complex scenes and perform their 3D reconstruction is our first choice for the continuation of the presented research.

Since the application of neural networks in semantic segmentation and object classification was justified by the reported accuracy in related research, training a neural network for general-purpose component segmentation and matching is an interesting subject for future research. In order to apply a neural network for solving the problem defined in this paper, a neural network should be trained using various datasets in order to learn a generic criterion for possible semantic association between model components. After being trained, the network would be applied for detection and association of the components of another unannotated dataset.

In this paper, a method for semantic association of the object components with the relatively small number of the annotated representative objects was proposed. The method was based on construction of the component association graph, whose nodes represented the object segments. Segments were approximated by convex polyhedrons. The CAG edges were assigned a semantic correspondence likelihood measure between the segments, which took into consideration both segments and their neighborhood similarity. The final association was performed by one of the three proposed methods based on the component association graph. The evaluation of the proposed approach on a subset of the ShapeNet Part benchmark dataset yielded interesting results, which indicated that with a rather small number of annotated representatives, the identification of a target component with accuracy measured by IoU greater than 0.6 could be achieved in only a few seconds.

Author Contributions: Conceptualization, R.C.; data curation, P.D.; formal analysis, P.D. and R.C.; funding acquisition, R.C.; investigation, P.D. and R.C.; methodology, P.D. and R.C.; project administration, R.C.; software, P.D., I.V., and R.C.; supervision, R.C.; validation, R.C.; visualization, R.C.; writing, original draft, P.D. and R.C.; writing, review and editing, P.D. and R.C. All authors have read and agreed to the published version of the manuscript.

Funding: This work has been fully supported by the Croatian Science Foundation under the project number IP-2014-09-3155.

Acknowledgments: We are especially grateful to Li Yi from Stanford University for kindly providing us the mapping between ShapeNet dataset meshes and point cloud model names and helping us with other dilemmas regarding ShapeNet.

Conflicts of Interest: The authors declare no conflict of interest. The funders had no role in the design of the study; in the collection, analyses, or interpretation of data; in the writing of the manuscript; nor in the decision to publish the results.

Appendix A. Segment Merging

The criterion used to select the candidates for merging is based on proximity of two segments and convexity of their union. Two segments C_i and C_j satisfy the proximity criterion if:

$$\min(\mu_{ij}, \mu_{ji}) \geq -\tau_{prox,1},$$

where $\tau_{prox,1} = 0.2$ is an experimentally determined threshold. The convexity of the union of two segments is evaluated by computing its convex hull and counting the outlier points, i.e., points that do not lie on the convex hull within a predefined threshold. The outlier percentage is defined by:

$$\eta(C_i, C_j) = \frac{|C_i \cup C_j| - |inliers(C_i \cup C_j, CH(C_i \cup C_j))|}{\min(|C_i|, |C_j|)},$$

where $|X|$ denotes the cardinality of a set X, $CH(X)$ denotes the convex hull of a set X, and $inliers(X, Y)$ denotes the inlier set. The inlier set contains all points $p \in X$ that are close to the surface of a convex polyhedron Y. Furthermore, $p \in inliers(X, Y)$ only if the local surface in the vicinity of p is similarly oriented as the surface of Y in the vicinity of p. This is evaluated by the following criterion based on the point-to-plane distance between p and the tangential plane of Y with the normal parallel to the local surface normal of p:

$$\max_{p' \in V_Y} (n_p^T(p' - p)) \leq \tau_{prox,2},$$

where n_p is the local surface normal of p, V_Y is the set of vertices of Y, and $\tau_{prox,2}$ is an experimentally determined threshold. In the experiments reported in Section 7, $\tau_{prox,2}$ was set to 5% of the radius of the model's bounding sphere, where an approximate minimum bounding sphere was computed by the algorithm proposed in [35]. If $\eta(C_i, C_j) < \tau_{new}$, then a new segment is created from the union of C_i and C_j. If $\eta(C_i, C_j) < \tau_{merge} < \tau_{new}$, then the new segment replaces C_i and C_j, i.e., the two original segments are rejected. The values of thresholds τ_{new} and τ_{merge} used in the experiments reported in Section 7 were 0.2 and 0.05, respectively. This merging process was applied not only to the original segments, but also to the new segments created by merging the original segments. In that case, the candidates for merging were always one original and one new segment. If one candidate was an original segment C_i and another was a new segment composed of original segments C_j, then the proximity and convexity criterion were evaluated for each pair of segments (C_i, C_j).

References

1. Abdo, N.; Kretzschmar, H.; Spinello, L.; Stachniss, C. Learning manipulation actions from a few demonstrations. In Proceedings of the 2013 IEEE International Conference on Robotics and Automation, Karlsruhe, Germany, 6–10 May 2013; pp. 1268–1275. [CrossRef]
2. Pastor, P.; Kalakrishnan, M.; Chitta, S.; Theodorou, E.; Schaal, S. Skill learning and task outcome prediction for manipulation. In Proceedings of the 2011 IEEE International Conference on Robotics and Automation, Shanghai, China, 9–13 May 2011; pp. 3828–3834. [CrossRef]

3. Nyarko, E.; Vidović, I.; Radočaj, K.; Cupec, R. A nearest neighbor approach for fruit recognition in RGB-D images based on detection of convex surfaces. *Expert Syst. Appl.* **2018**, *114*, 454–466. [CrossRef]
4. Yi, L.; Shao, L.; Savva, M.; Huang, H.; Zhou, Y.; Wang, Q.; Graham, B.; Engelcke, M.; Klokov, R.; Lempitsky, V.; et al. Large-Scale 3D Shape Reconstruction and Segmentation from ShapeNet Core55. *arXiv* **2017**, arXiv:1710.06104.
5. Chang, A.X.; Funkhouser, T.; Guibas, L.; Hanrahan, P.; Huang, Q.; Li, Z.; Savarese, S.; Savva, M.; Song, S.; Su, H.; et al. *ShapeNet: An Information-Rich 3D Model Repository*; Technical Report arXiv:1512.03012 [cs.GR]; Stanford University—Princeton University—Toyota Technological Institute at Chicago: Chicago, IL, USA, 2015.
6. Li, Y.; Bu, R.; Sun, M.; Wu, W.; Di, X.; Chen, B. PointCNN: Convolution On $\mathcal{X}$-Transformed Points. *arXiv* **2018**, arXiv:1801.07791.
7. Garcia-Garcia, A.; Orts-Escolano, S.; Oprea, S.; Villena-Martinez, V.; Garcia-Rodriguez, J. A Review on Deep Learning Techniques Applied to Semantic Segmentation. *arXiv* **2017**, arXiv:1704.06857.
8. Yi, L.; Guibas, L.; Kim, V.G.; Ceylan, D.; Shen, I.C.; Yan, M.; Su, H.; Lu, C.; Huang, Q.; Sheffer, A. A scalable active framework for region annotation in 3D shape collections. *ACM Trans. Graph.* **2016**, *35*, 1–12. [CrossRef]
9. Yi, L.; Guibas, L.; Hertzmann, A.; Kim, V.G.; Su, H.; Yumer, E. Learning Hierarchical Shape Segmentation and Labeling from Online Repositories. *ACM Trans. Graph.* **2017**, *36*, 1–12. [CrossRef]
10. Kalogerakis, E.; Hertzmann, A.; Singh, K. Learning 3D mesh segmentation and labeling. *ACM Trans. Graph.* **2010**, *29*, 1–12. [CrossRef]
11. Sun, X. Semantic Annotation of 3D Architecture Models Based on the Geometric Structure Characteristics. In Proceedings of the 2018 26th International Conference on Geoinformatics, Kunming, China, 28–30 June 2018; pp. 1–6. [CrossRef]
12. Wang, N.; Zhang, Y.; Li, Z.; Fu, Y.; Liu, W.; Jiang, Y.G. Pixel2Mesh: Generating 3D Mesh Models from Single RGB Images. In *Computer Vision—ECCV 2018*; Ferrari, V., Hebert, M., Sminchisescu, C., Weiss, Y., Eds.; Lecture Notes in Computer Science; Springer International Publishing: Munich, Germany, 2018; Volume 11215, pp. 55–71. [CrossRef]
13. Groueix, T.; Fisher, M.; Kim, V.G.; Russell, B.C.; Aubry, M. AtlasNet: A Papier-Mache Approach to Learning 3D Surface Generation. *arXiv* **2018**, arXiv:1802.05384.
14. Mescheder, L.; Oechsle, M.; Niemeyer, M.; Nowozin, S.; Geiger, A. Occupancy Networks: Learning 3D Reconstruction in Function Space. In Proceedings of the IEEE Conference on Computer Vision and Pattern Recognition (CVPR), Long Beach, CA, USA, 16–20 June 2019.
15. Yi, L.; Zhao, W.; Wang, H.; Sung, M.; Guibas, L. GSPN: Generative Shape Proposal Network for 3D Instance Segmentation in Point Cloud. *arXiv* **2018**, arXiv:1812.03320.
16. Cupec, R.; Vidović, I.; Filko, D.; Durovic, P. Object Recognition Based on Convex Hull Alignment. *Pattern Recognit.* **2020**, [CrossRef]
17. Cupec, R.; Filko, D.; Nyarko, E.K. Segmentation of depth images into objects based on local and global convexity. In Proceedings of the 2017 European Conference on Mobile Robots (ECMR), Paris, France, 6–8 September 2017; p. 1–7. [CrossRef]
18. Durovic, P.; Filipovic, M.; Cupec, R. Alignment of Similar Shapes Based on their Convex Hulls for 3D Object Classification. In Proceedings of the 2018 IEEE International Conference on Robotics and Biomimetics (ROBIO), Kuala Lumpur, Malaysia, 12–15 December 2018; pp. 1586–1593. [CrossRef]
19. Cupec, R.; Durovic, P. Volume Net: Flexible Model for Shape Classes. In Proceedings of the 2018 IEEE International Conference on Robotics and Biomimetics (ROBIO), Kuala Lumpur, Malaysia, 12–15 December 2018; pp. 248–255. [CrossRef]
20. Cupec, R.; Filko, D.; Durovic, P. Segmentation of Depth Images into Objects Based on Polyhedral Shape Class Model. In Proceedings of the 2019 European Conference on Mobile Robots (ECMR), Prague, Czech Republic, 4–6 September 2019; pp. 1–8. [CrossRef]
21. Wohlkinger, W.; Aldoma, A.; Rusu, R.B.; Vincze, M. 3dnet: Large-scale object class recognition from cad models. In Proceedings of the 2012 IEEE International Conference on Robotics and Automation, Saint Paul, MN, USA, 14–18 May 2012; pp. 5384–5391.
22. Yi, L.; Su, H.; Guo, X.; Guibas, L. SyncSpecCNN: Synchronized Spectral CNN for 3D Shape Segmentation. In Proceedings of the 2017 IEEE Conference on Computer Vision and Pattern Recognition (CVPR), Honolulu, HI, USA, 21–26 July 2017; pp. 6584–6592. [CrossRef]

23. Klokov, R.; Lempitsky, V. Escape from Cells: Deep Kd-Networks for the Recognition of 3D Point Cloud Models. In Proceedings of the 2017 IEEE International Conference on Computer Vision (ICCV), Honolulu, HI, USA, 21–26 July 2017; pp. 863–872. [CrossRef]
24. Graham, B.; Engelcke, M.; van der Maaten, L. 3D Semantic Segmentation with Submanifold Sparse Convolutional Networks. In Proceedings of the 2018 IEEE/CVF Conference on Computer Vision and Pattern Recognition, Salt Lake City, UT, USA, 18–22 June 2018; pp. 9224–9232. [CrossRef]
25. Xu, Y.; Fan, T.; Xu, M.; Zeng, L.; Qiao, Y. SpiderCNN: Deep Learning on Point Sets with Parameterized Convolutional Filters. In *Computer Vision—ECCV 2018*; Ferrari, V., Hebert, M., Sminchisescu, C., Weiss, Y., Eds.; Series Title: Lecture Notes in Computer Science; Springer International Publishing: Munich, Germany, 2018; Volume 11212, pp. 90–105. [CrossRef]
26. Li, J.; Chen, B.M.; Lee, G.H. SO-Net: Self-Organizing Network for Point Cloud Analysis. In Proceedings of the 2018 IEEE/CVF Conference on Computer Vision and Pattern Recognition, Salt Lake City, UT, USA, 18–22 June 2018; pp. 9397–9406. [CrossRef]
27. Atzmon, M.; Maron, H.; Lipman, Y. Point convolutional neural networks by extension operators. *ACM Trans. Graph.* **2018**, *37*, 1–12. [CrossRef]
28. Shen, Y.; Feng, C.; Yang, Y.; Tian, D. Mining Point Cloud Local Structures by Kernel Correlation and Graph Pooling. In Proceedings of the 2018 IEEE/CVF Conference on Computer Vision and Pattern Recognition, Salt Lake City, UT, USA, 18–22 June 2018; pp. 4548–4557. [CrossRef]
29. Ben-Shabat, Y.; Lindenbaum, M.; Fischer, A. 3D Point Cloud Classification and Segmentation using 3D Modified Fisher Vector Representation for Convolutional Neural Networks. *arXiv* **2017**, arXiv:1711.08241.
30. Huang, Q.; Wang, W.; Neumann, U. Recurrent Slice Networks for 3D Segmentation of Point Clouds. In Proceedings of the 2018 IEEE/CVF Conference on Computer Vision and Pattern Recognition, Salt Lake City, UT, USA, 18–22 June 2018; pp. 2626–2635. [CrossRef]
31. Wang, Y.; Sun, Y.; Liu, Z.; Sarma, S.E.; Bronstein, M.M.; Solomon, J.M. Dynamic Graph CNN for Learning on Point Clouds. *arXiv* **2018**, arXiv:1801.07829.
32. Qi, C.R.; Su, H.; Mo, K.; Guibas, L.J. Pointnet: Deep learning on point sets for 3d classification and segmentation. In Proceedings of the 2017 IEEE Conference on Computer Vision and Pattern Recognition (CVPR), Honolulu, HI, USA, 21–26 July 2017; Volume 1, p. 4.
33. Qi, C.R.; Yi, L.; Su, H.; Guibas, L.J. PointNet++: Deep Hierarchical Feature Learning on Point Sets in a Metric Space. *arXiv* **2017**, arXiv:1706.02413.
34. Wang, W.; Yu, R.; Huang, Q.; Neumann, U. SGPN: Similarity Group Proposal Network for 3D Point Cloud Instance Segmentation. *arXiv* **2019**, arXiv:1711.08588.
35. Ritter, J. *An Efficient Bounding Sphere*; Glassner, A.S., Ed.; Graphics Gems, Academic Press Professional, Inc.: Ithaca, NY, USA, 1990; pp. 301–303. [CrossRef]

© 2020 by the authors. Licensee MDPI, Basel, Switzerland. This article is an open access article distributed under the terms and conditions of the Creative Commons Attribution (CC BY) license (http://creativecommons.org/licenses/by/4.0/).

Article

Method for Volume of Irregular Shape Pellets Estimation Using 2D Imaging Measurement

Andrius Laucka [1], Darius Andriukaitis [1,*], Algimantas Valinevicius [1], Dangirutis Navikas [1], Mindaugas Zilys [1], Vytautas Markevicius [1], Dardan Klimenta [2], Roman Sotner [3] and Jan Jerabek [4]

[1] Department of Electronics Engineering, Kaunas University of Technology, Studentu St.50-438, LT-51368 Kaunas, Lithuania; andrius.laucka@ktu.edu (A.L.); algimantas.valinevicius@ktu.lt (A.V.); dangirutis.navikas@ktu.lt (D.N.); mindaugas.zilys@ktu.lt (M.Z.); vytautas.markevicius@ktu.lt (V.M.)
[2] Faculty of Technical Sciences Kosovska Mitrovica, University of Pristina in Kosovska Mitrovica, Kneza Milosa St. 7, RS-38220 Kosovska Mitrovica, Serbia; dardan.klimenta@pr.ac.rs
[3] Department of Radio Electronics, Brno University of Technology, Technicka 3082/12, 616 00 Brno, Czech Republic; sotner@feec.vutbr.cz
[4] Department of Telecommunications, Brno University of Technology, Technicka 3082/12, 616 00 Brno, Czech Republic; jerabekj@feec.vutbr.cz
* Correspondence: darius.andriukaitis@ktu.lt

Received: 19 March 2020; Accepted: 9 April 2020; Published: 11 April 2020

Abstract: Growing population and decreasing amount of cultivated land conditions the increase of fertilizer demand. With the advancements of computerized equipment, more complex methods can be used for solving complex mathematical problems. In the fertilizer industry, the granulometric composition of products matters as much as the quality of production of chemical composition products. The shape and size of pellets determines their distribution over cultivated land areas. The effective distance of field spreading is directly related to the size and shape parameters of a pellet. Therefore, the monitoring of production in production lines is essential. The standard direct methods of the monitoring and control of granulometric composition requires too much time and human resources. These factors can be eliminated by using imaging measuring methods that have a variety of benefits, but require additional research in order to assure and determine the compliance of real-time results with results of the control equipment. One of the fastest, most flexible and largest amount of data providing methods is the processing and analysis of digital images. However, then we face the issue of the suitability of 2D images to be used for the evaluation of granulometric compositions, where processing of digital images provides only two dimensions of a pellet: length and width. This study proposes a method of evaluating an irregular pellet. After experimental research we determined < 2% of discrepancy when compared to the real volume of a pellet.

Keywords: image processing; fertilizers; distribution; monitoring

1. Introduction

The rapid application of modern technologies is evident in many various industries. The increasing performance of computer equipment and the wider realization of various theoretical mathematical calculations expands the possibilities of applications of equipment. The largest volumes of production are conditioned by application of cutting edge technical and scientific solutions in various fields of industry. The chemistry industry is no exception. The increasing demand of food supplies and scarcity of cultivated land is a reason for the wider application of fertilizers. Rational fertilization of cultivated land and optimal distribution across the surface of soil or insertion into soil results in better harvests. Therefore, there are specific requirements for the size and shape of fertilizer pellets. However, it is a huge task for manufacturers to assure the quality of fertilizers. As the fertilizer pellets are quite small (2–4 mm in diameter), it is difficult to evaluate their quality by using direct methods of measuring.

With increasing productivity, the best tools for quality assurance are indirect methods of measuring that allow real-time evaluation of the quality of pellets [1–13]. The technique of processing digital images provides results that are very similar to the results of control equipment based on direct measurements [14–19]. A variety of image enhancement methods assure reliability of results [20]. A high frequency of discretization of production monitoring allows for a quick reaction to any production process deviations that manifest in changes of production parameters. Applied numerical intellect solutions simplify the preparation of equipment for work [21,22].

One of the most important qualitative parameters used for the evaluation of pellets is their distribution according to size [23]. However, other parameters are not less important, such as: distribution according to roundness, elongation, compactness, uniformity index of pellets and so on [24]. According to the international standards of sampling and the evaluation of samples [25–27], the manufacturer strictly defines qualitative parameters related to other physical parameters. Round pellets and pellets that are categorized by nondispersive size distribution are more likely to be more evenly distributed during fertilization using an automatic spreader [28,29].

Optimal spreading conditions are achieved with higher bulk density, because it creates further throwing distance of the pellets. Various fertilization quality assessment systems are installed depending on the type of spreader used [30]. However, it must be combined with optimal distribution interval of pellets—ideally, distribution interval deviates by ±0.8 mm (width) from the average [31]. Additionally, the more the shape of the pellet is similar to an ideal circle, the longer the distance of its flight, because of its aerodynamic properties. Hardness of the pellet prevents the disks of the pellet spreader from breaking it apart [32].

One of the relevant problems related to using indirect measuring methods is the matching of the results to those from using direct methods. It is especially difficult to evaluate the pellets of irregular shape. According to the literature analysis, many various methods are used for scanning of pellets as well as recalculation of results. The main purpose of this article is to propose a fast method for evaluating the quality parameters of fertilizer pellets, which would ensure continuous quality control. This can only be achieved through indirect measurements. Digital image analysis allows for versatile and fast output analysis. However, there are problems with the evaluation of irregularly shaped pellets in two-dimensional images. This increases the discrepancy between the actual results and the results of the indirect measurement. In this work, the proposed method of evaluation of fertilizer pellets of irregular shape is via processing and analysis of digital images. The obtained results of the proposed measurement method can be applied in the fertilizer production lines for continuous monitoring of production quality. In Section 2 (Materials and Methods), methods of pellet volume evaluation, when their scanning is carried out using imaging measurement techniques, are analyzed. In the second part of the work (Results, Subsection—Evaluation of pellet's shape) experimental tests are carried out in order to evaluate the production, which is the studied product. Shape characteristics of the pellets (studied product) and related problems are determined. Further on, the applied method of pellet volume calculation, where pellets are divided into layers of least possible height, is analyzed (Subsection—Division of pellet volume into layers). General volume is calculated by integrating the layers of pellet. Based on the received results, an improved method of the calculation of pellet volume was proposed, namely, evaluation of layer using geometric form—ellipse (Subsection—Approximation of pellet cross-section layers using ellipse). In Section 4 (Discussion) information about the improved method of pellet volume evaluation is presented. Meanwhile, conclusions are given in the final section.

2. Materials and Methods

The amount of data received from analysis of the digital image processing is limited for determining the distribution of pellets. Depending on the material being analyzed, pellets can be of irregular shapes, and any approximation of such pellets may largely differ from their real surface.

Two-dimensional (2D) image information always provides only two parameters of pellet size: length and width. Therefore, the precision of evaluation of its volume depends on the level of matching between the approximation model and real surface of a pellet. The contours of pellets are usually approximated using geometric forms that are most widely used in calculations, such as circles, ellipses or best fit rectangle (smallest area limited by rectangle that can define the shape of pellet) [33]. These models are used for defining the contour of a pellet, as well as drawing the largest possible shape (limited by the perimeter of a pellet) inside of a contour of a pellet (Figure 1).

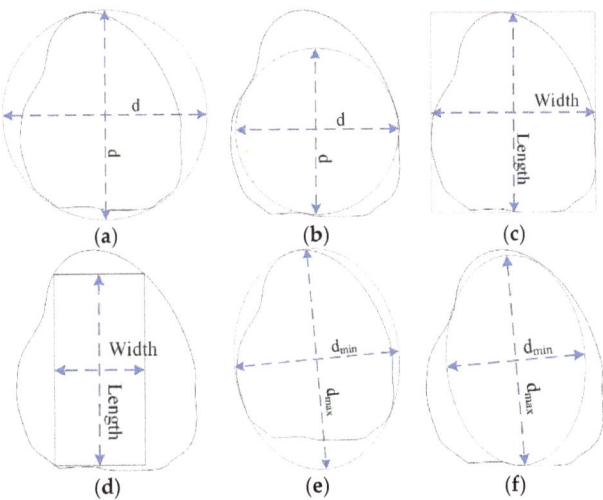

Figure 1. Models of pellet approximation: (a,b)—pellet is approximated using a circle drawn around it and a circle drawn inside of it; (c,d)—pellet is approximated using best fit rectangle drawn around it and rectangle drawn inside of it; (e,f)—pellet is approximated using a ellipse drawn around it and a ellipse drawn inside of it.

Such forms of approximation are characterized by limitedness, as in the case of an irregular pellet shape the area limited by its contour can be up to two times larger than the real area of the pellet (Figure 2a). This approximation form is the reason for discrepancy between measured volume and real measurement results. To account for such discrepancies, a model of the area (matching real area of a pellet) limited by the regular geometric shape is used (Figure 2b).

Such methods of pellet shape evaluation are used with silhouette contour approximation using ellipses [34], because this geometric shape allows for better evaluation of roundness of the pellet, when compared to a circle shape. The ellipse contour provides information about longitudinal and transverse dimensions. This method of measurement is similar to the sieve method. In many laboratories, sets of sieves are used as a control equipment of examination. In this case, when distribution of pellets must be determined, the main parameter is the width of the ellipse that limits the area of the pellet. The width of the ellipse is the same as the diameter of the holes of the sieve. Matching between the holes of a sieve and the approximated form is presented in Figure 3.

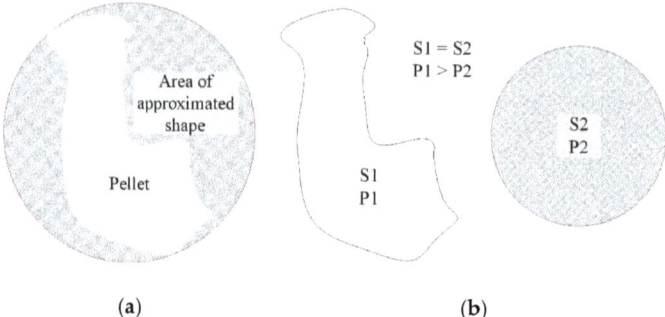

Figure 2. Connection between pellet and the shape that is being approximated: (**a**) area of discrepancy between real area of irregular pellet and circle shaped approximation is marked by the shaded area; (**b**) left side—real area limited by pellet (S1) and perimeter (P1); right side—circle shape formed from the area of pellet.

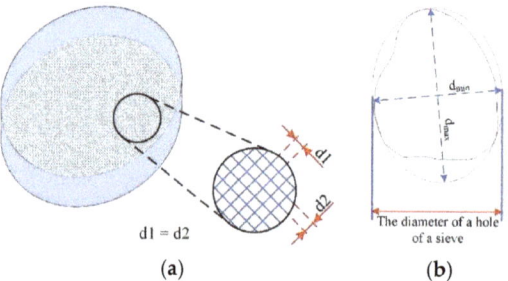

Figure 3. Similarity of results of pellet approximation model and control equipment. (**a**) Control measuring equipment consisting of a set of sieves with different diameter holes; (**b**) the width of pellet approximation shape matches the diameter of a hole of a sieve.

The sieve method used for the measuring of pellets is not a rational choice for evaluation of parameters of their shape. This is related to the working principle of sieves, which only allows for evaluation of two out of three dimensions of pellet, as the pellet falls through sieve when two of its three transverse dimensions are smaller than the diameter of holes of the sieves. In such a case, a third diameter as well as shape of the pellet is left unmeasured. The precision of measuring is also limited by the number of sieves with holes of different diameters. The technique of processing digital images expands the measuring possibilities, while more precise mathematical models used for pellet approximation conditions the improvement of precision and reliability. The correction of the final distribution of the pellets is used as an alternative, e.g., an evaluation carried out only with pellets which are shaped very similarly to the regular shape of a pellet [35].

During the indirect measuring, based on digital image processing, various geometric shapes are used for the approximation of pellets, according to the area limited by their silhouette. One of most common geometric shapes is volume limited by ellipsoid Equation (1) (according to Figure 1e) [36]:

$$V = \frac{4}{3}\pi r_1 r_2 \frac{d_2}{2}, \tag{1}$$

where r_1, r_2—radius of ellipsoid; d_2—length of ellipsoid.

The analyzed scan of the pellet image is two dimensional, therefore in calculation we additionally use the width of the ellipsoid, i.e., instead of one radius of the ellipsoid, which is equal to the depth of image, the other radius is used in Equation (2):

$$V = \frac{4}{3}\pi r_2 r_2 \frac{d_2}{2} = \frac{4}{3}\pi r_2^2 \frac{d_2}{2}, \qquad (2)$$

where r_2—radius of ellipsoid; d_2—length of ellipsoid (ellipsoid scheme in Figure 4).

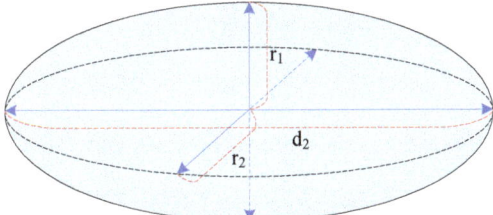

Figure 4. The ellipse model.

When pellets are analyzed in terms of volume distribution using Equation (2), each pellet is evaluated as a shape of the ideal ellipse. Such an evaluation will be correct only in specific cases, when the product is characterized by a high index of roundness and smoothness of surfaces. These parameters are determined by chemical composition of a pellet. It determines their fragility and adhesion of mixture. Therefore, when working with irregularly shaped pellets, the method of ellipsoid volume evaluation causes a discrepancy between the results and the real volume of such pellet.

3. Results

The reliability of the evaluation of pellet volume depends on the shape of a pellet. An evenly shaped, not fractured pellet that is resistant to mechanical impact can be evaluated precisely, using approximation of a circle shape. However, in reality we see many irregularly shaped pellets. Therefore, their evaluation using 2D imaging decreases the reliability of the method. Discrepancy is introduced with irregularities of the pellet's surface. For developing a method of the evaluation of volume of irregularly shaped pellets, primary analysis of production is necessary, because the shape of surface irregularities must be evaluated.

3.1. Evaluation of Shape of Pellets

Depending on the chemical composition of fertilizers, they can be characterized by anisotropic shape. Therefore, ordinary methods of pellet volume by approximation using different geometric shapes introduces discrepancies. For this reason, with any method of pellet volume evaluation, the product's distribution according to roundness should be known. The distribution of roundness of monoammonium phosphate pellets according to ratio between diagonals of area limited by pellet is provided in Figure 5. During the measurement, 100 different samples taken during the production were analyzed. The graph shows the average of the measurements, and standard squared deviation of measurement < 4%.

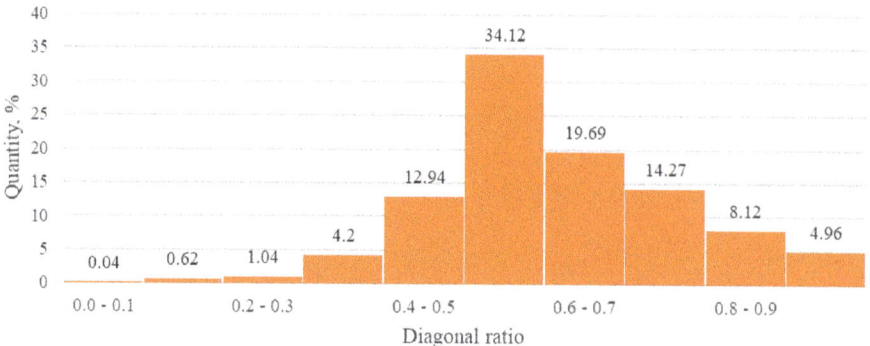

Figure 5. Distribution of monoammonium pellets according to roundness (ratio of diagonals of area limited by pellet's silhouette).

After evaluating the results of measuring of roundness of the studied pellets, the assumption can be made that irregularly shaped, elongated pellets are most common throughout all production. Since techniques of processing two-dimensional images cannot take the depth of the image into account, the diameters of pellets, perpendicular to their longitudinal axis, must be thoroughly evaluated. During the examination of shape evaluation, ~1000 pellets were analyzed. A quantity of 10–20 pellets were randomly chosen from each roundness interval of different samples. Every pellet was scanned twice—an image of the largest area limited by the pellet, and an image of the pellet rotated by 90° (principle of pellet evaluation is presented in Figure 6). Results of pellet measuring are provided in Figure 7.

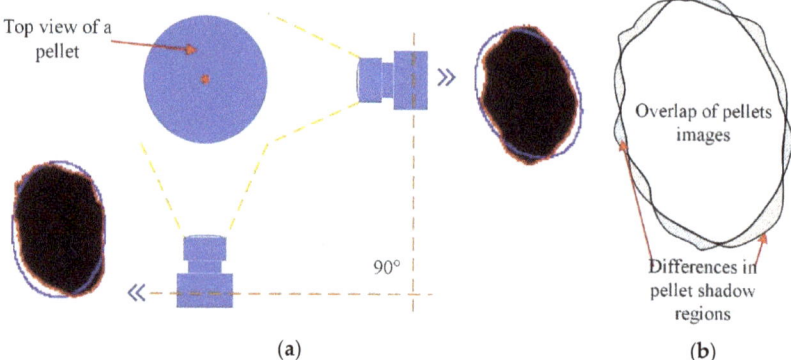

Figure 6. (a) The pellet is scanned using a 90° angle between the camera's field of view and the longitudinal axis of the pellet, when we receive the largest area limited by the pellet and image rotated by 90°; (b) evaluation of discrepancy of pellet overlapping area according to averages of areas.

Results received from analysis of monoammonium phosphate pellets indicate that discrepancy of shape of pellets increases with increasing size of a pellet. However, ≤ 95% of all pellets of the material with which we are working are distributed in the interval from 0.2 mm to 1.0 mm. Within these boundaries, roundness of pellets according to ration of diagonals is distributed within boundaries from 0.4 to 0.8. The discrepancy among areas limited by different contours produced according to longitudinal axis that is in the mentioned intervals is < 3%. Because of the contamination of equipment, pellets tend to become smaller during production and it is sufficient to evaluate their shape from one position.

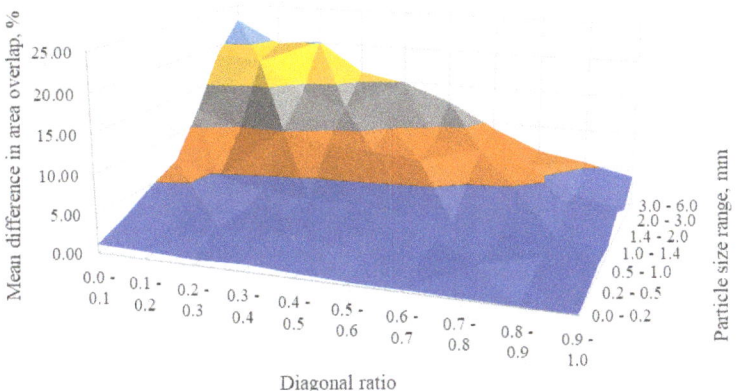

Figure 7. Average discrepancy of area limited by the pellet's silhouette when pellets are evaluated according to their size and roundness.

3.2. Estimation of Pellet Volume by Dividing Them into Circular Layers

Based on experiments of pellet analysis, it can be stated that sufficient precision can be achieved when the pellet's volume is evaluated according to area limited by the pellet's silhouette. After evaluating examination results it was determined that only in specific cases, when dealing with a pellet's shape that is very close to ideal, is the ellipsoid volume formula Equation (2) suitable for calculation of pellet volume. Since roundness index of analyzed material is not very high (evidenced by experimental examination, Figure 4), this mathematical expression does not match the real volume limited by pellet. According to the literature [37], the reduction of error of measurements is affected by the division of the analyzed ellipsoid into layers, i.e., dividing its area into narrow regions that are perpendicular to longitudinal axis of ellipse.

The area limited by the pellet's silhouette is divided into layers of least height. This height is equal to the height of one point of the image. Using this method, the area of circle, limiting every layer, is calculated [37,38] with Equation (3):

$$S_i = \pi r^2 = \pi \left(\frac{\Delta y}{2}\right)^2, \qquad (3)$$

where S_i—area of circle that limits the layer; Δy—the widthof the ellipse layer (particle scheme in Figure 8).

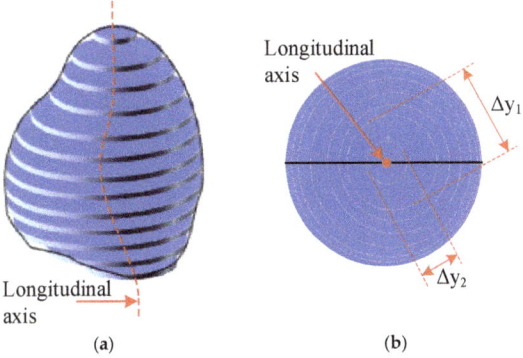

Figure 8. The layering of an ellipse: (**a**) side view of an ellipse model; (**b**) top view of ellipse model.

The measuring method studied by Rashidi and Gholami [37] is based on model of the ideal ellipse, because the ellipsoid consists of sum of volumes of circles Equations (4) and (5):

$$V_i = S_i \Delta x, \tag{4}$$

$$V = \sum_{i=1}^{n} V_i, \tag{5}$$

where S_i—area of circle limiting the layer; V_i—volume of circle limiting the layer; V—volume limited by ellipsoid.

During the examination, the formed models of the regular shaped pellets were scanned using video camera. Pellets of ellipse shape were formed according to measurement results presented in Figure 7—most pellets are elongated. At the beginning of measuring, the calibration of the experimentation equipment was carried out by evaluating the ratio coefficient between points in the image and real size of a pellet—for recalculation of results into SI (metric) system units. In the later stages, after binarization and threshold segmentation operations were carried out in scanned images, the area of the analyzed pellet was distinguished. The general volume of the pellet was calculated by adding up volumes of the separates layers (perpendicular to longitudinal axis of the pellet), whose height is equal to one point of image. The algorithm of the calculation method is presented in Figure 9. Convexity of the was evaluated experimentally. During the examination of pellet models in course of the experiment, it was determined that this threshold value was 0.8, while there were additional calculations carried out for determining the convexity of a pellet.

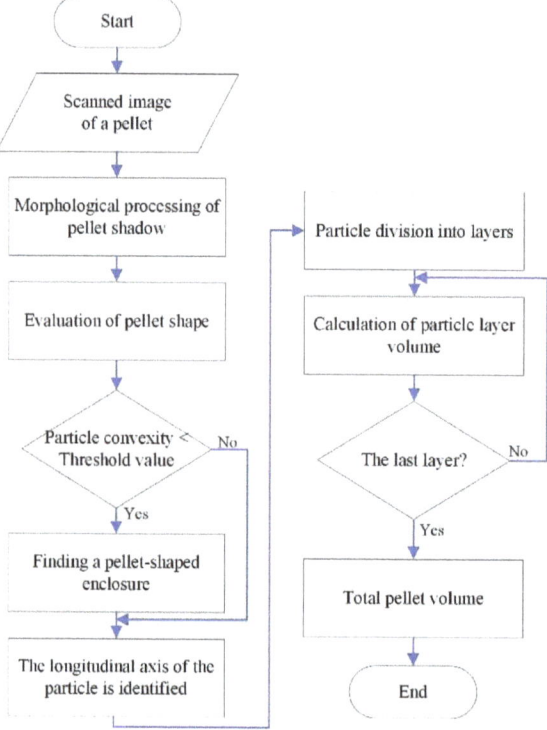

Figure 9. Algorithm for calculation of pellet volume (pellets divided into layers).

Additionally, the limited area of the pellet, as ellipsoid, was evaluated during the examination. The received results of the measurements and calculations are presented in Table 1. The results of measuring were compared to data received using the water displacement method. This method is considered to be benchmark method of measuring. Contours of the pellets were also approximated using circular geometric form. The volume of the granules was estimated using the circular approximation method presented in the Figure 1a. Results are presented in graphs of Figure 10.

Table 1. Calculation of volume of pellet model (pellet was divided into circular layers).

Sample no.	Water Displacement Method			Method for Dividing Pellets into Circular Layers				$\Delta V2$, cm^3	$\Delta V3$, cm^3
	Length of pellet, mm	Width of pellet, mm	Volume V1, cm^3	Length of pellet, mm	Width of pellet, mm	Volume V2, cm^3	Volume V3 [1], cm^3	$|V1-V2|$, cm^3	$|V1-V3|$, cm^3
1	80.4	18.0	13.8	81.11	18.37	14.11	14.33	0.31	0.52
2	68.3	17.6	13.4	69.38	17.86	13.18	11.58	0.22	1.82
3	49.1	20.7	13.6	49.65	21.09	13.36	11.56	0.24	2.04
4	53.4	20.6	13.8	53.90	21.26	13.85	12.75	0.05	1.05
5	56.1	20.8	14.0	56.46	21.09	14.00	13.14	0.01	0.86
6	61.4	18.6	13.2	62.41	19.22	12.93	12.07	0.27	1.13
7	74.0	18.2	14.0	74.14	18.54	14.01	13.34	0.01	0.66
8	74.7	17.8	14.0	75.16	18.03	14.01	12.79	0.01	1.21
9	66.9	19.1	14.4	67.17	19.22	14.20	12.99	0.20	1.41
10	64.8	18.3	13.8	65.13	18.71	13.58	11.93	0.22	1.87
							MAE, cm^3	1.258	0.153
							SSE	6.872	1.711
							σ^2	0.687	0.171
							RMSE	1.351	0.191

[1] Volume of ellipsoid (area of ellipse approximated using ellipse).

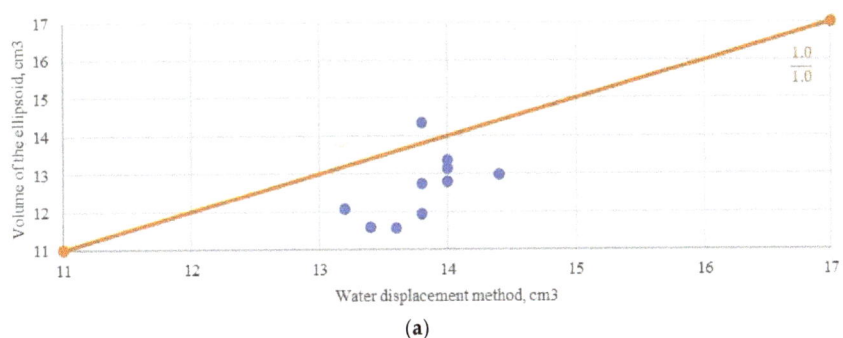

(a)

Figure 10. *Cont.*

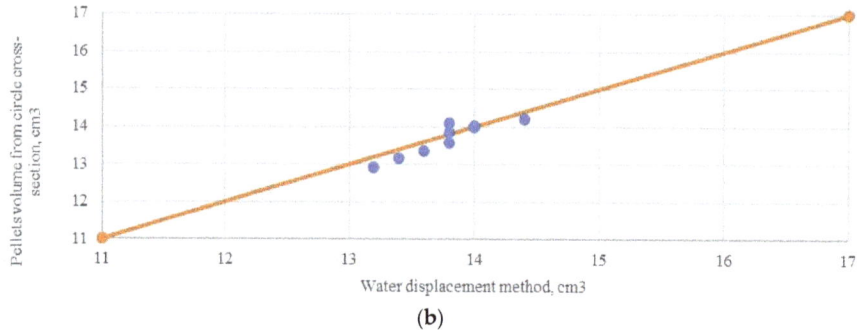

(b)

Figure 10. Measuring of volume of formed pellet model based on imaging method of processing of digital images: (**a**) ratio between results produced by approximation of ellipsoid volume method and water displacement method; (**b**) ratio between results produced by approximation of pellet volume by dividing pellet's volume into layers of circles and water displacement method.

After evaluation of results of the pellet volume measurements, presented in Figures 1 and 10, it can be stated that approximation using the ellipse limiting the pellet's area is not a reliable volume measuring method for evaluating the volume of the ellipsoid. In case of ellipsoid approximation, the average of absolute error of measurements is up to 1.258 cm^3, while σ = 0.687. When the volume of pellets model is calculated by dividing its area into layers of one point of the image and the limited area of layers is added up, more reliable measurement results are produced because of the lowest possible distance between the adjacent analyzed areas of cross-section. Respectively, the average of absolute error—0.153 cm^3 and σ = 0.171. To evaluate the results produced during the experiment, it can be said that applied method of pellet evaluation of analysis of two-dimensional image of a pellet by dividing it into limited areas of cross-section is reliable and assures low MRE (Mean Relative Error) Equation (6):

$$\delta = \frac{|0.15|}{|13.8|} \times 100\% \approx 1.109\%. \tag{6}$$

It was determined that the applied method of pellet shape evaluation is effective with regular geometric shape of a pellet, which rarely matches real irregularities limited by the area of an angled pellet. For evaluation of the shape, 50 samples of same material (monoammonium phosphate) were randomly taken at different point of time of production. Figure 11 provides two dimensional images of scanned irregularly shaped pellets together with statistical graphs of distribution of irregularly shaped pellets in a sample, according to pellet's roundness.

After evaluating the samples taken in course of the production, it was determined that the statistically most common irregularities of pellet shape were convexity, flatness of shape and various notches. Convexity of shape can be evaluated using the Equation (5) volume calculation formula, where different layers of pellet are treated as circles. However, in such cases each irregularity limited by the region of the pellet silhouette should be positioned in an area limited by $> \frac{3}{4}$ of the pellet cross-section (Figure 12). When the pellet is flat, evaluation of its volume according to circles, which the cross-section consists of, creates discrepancies of measurements.

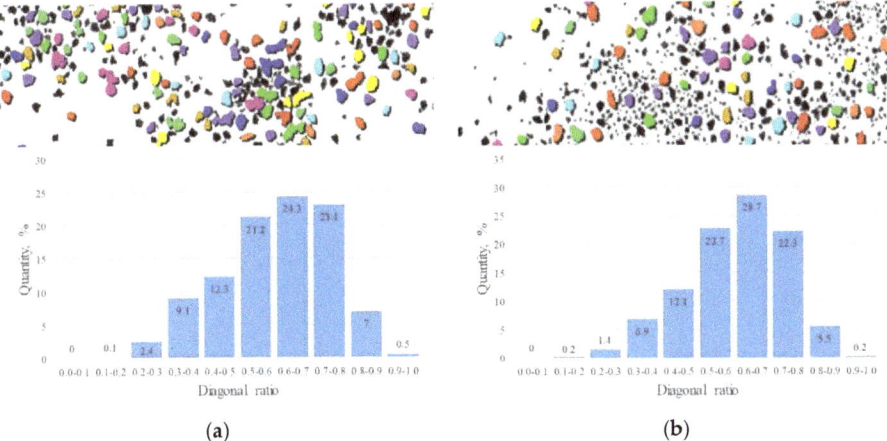

Figure 11. Analysis of example of pellets sample: (a) distribution of pellets of first sample according to roundness; (b) distribution of pellets of second sample according to roundness.

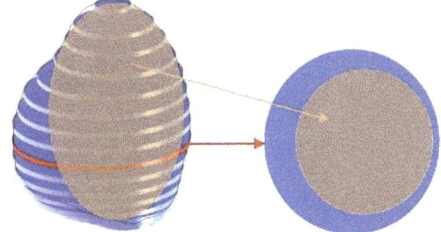

Figure 12. Largest possible ellipse, whose volume is limited by region of pellet, is drawn inside of a pellet.

The cross-section of flat pellets or pellets with notches with respect to longitudinal axis of pellet is similar to the area limited by ellipse. According to previously presented methods of pellet volume evaluation, where separate layers of pellets are evaluated as circles, it is likely that measurement discrepancy will appear between measurement results, when they are compared to data received using control equipment. In the case of evaluating irregularly shaped pellets, calculation of circle area in place of a notch in the pellet decreases the area of its cross-section and this directly affects the shrinking of pellet volume.

3.3. Estimation of Pellet Volume by Dividing Them into Elliptical Layers

For compensation of pellet volume calculation when the pellet is irregularly shaped, convexity of the area limited by the pellet was evaluated. After filling the notch in the pellet with its convex shape, its depth can be rationally evaluated. The ellipse geometric shape fills the cross-section area of a pellet more precisely. The three-dimensional model of irregular pellet presented in Figure 13.

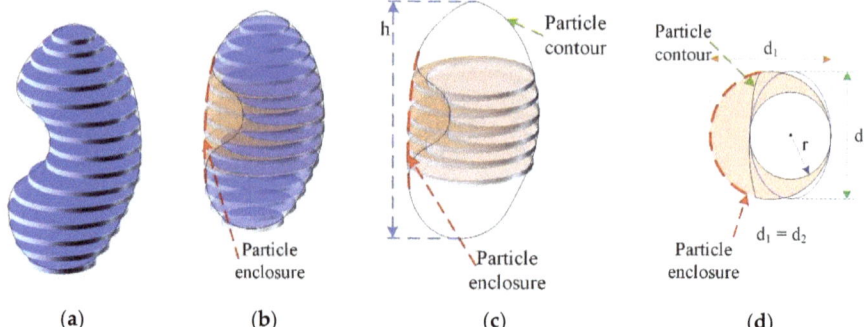

Figure 13. Three-dimensional (3D) model of irregular pellet: (**a**) the pellet's volume consists of the sum of layer circles; (**b**) shape of the regular pellet is limited by pellet envelope line; (**c**) discrepancy between irregular volume and volume limited by envelope line; (**d**) cross-section of pellet limited by its envelope line.

Based on the pellet model presented in Figure 13, an assumption can be made that the cross-section area of the pellet limited by the envelope line is better overlapped by an ellipse than a circle. The volume is calculated by using both contours of the pellet—the real and notched area limited by the envelope line. The pellets volume is divided into cross-section areas, whose height is equal to one point of the image and which are perpendicular to longitudinal axis; then the pellets volume is calculated by integrating (according to Figure 13) Equations (7) and (8):

$$S(h) = \pi r \frac{d_2}{2}, \tag{7}$$

$$V = \int_0^h S(h) dh, \tag{8}$$

where S(h)—area of ellipse limiting the layer; r—half of cross-section of real area limited by pellet; d_2—diameter of cross-section limited by pellets envelope line; h—length of a pellet, perpendicular to planes of layers that limit it; V—volume limited by pellet.

For testing the reliability of the measuring method evaluating irregularly shaped pellets, models of plasticine were made (Figure 14a). Formed models of the pellets were scanned in two directions, using video camera, with the pellets position covering the largest possible area in relation to position of the video camera and the pellet rotated by 90° from the largest area position. This allowed us to evaluate the reliability of the measuring method by analyzing irregular geometric shapes from different positions in space. Calibrated experimental equipment was used for measuring. After carrying out primary morphological operations in the scanned images, the contour limited by the model was identified (Figure 14b). One of more important objectives of the primary processing is elimination of very small pellets that are in the background of the image and are not affected by gravity. This task is carried out by applying evaluation of the background [39], which allows us to filter out unneeded objects [40].

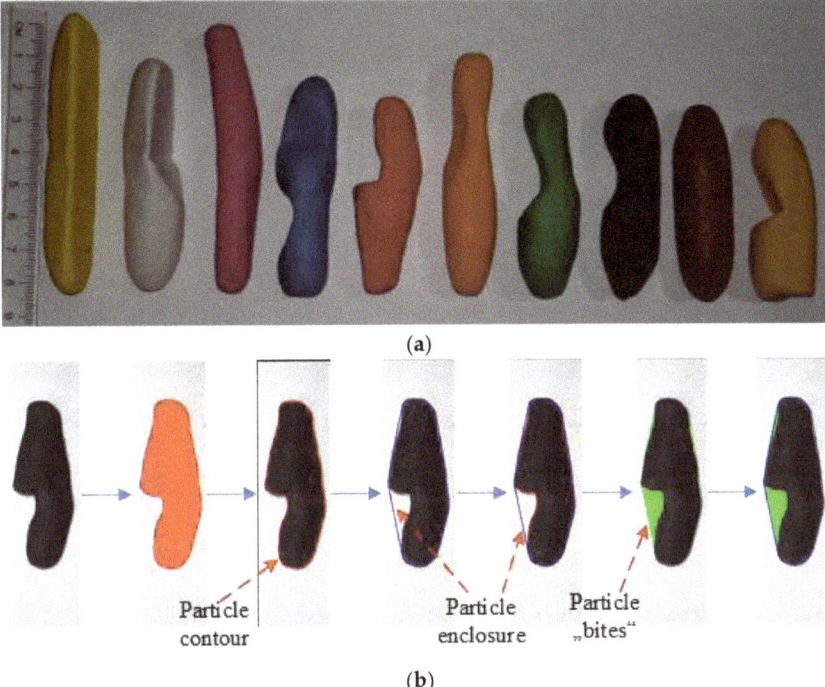

Figure 14. (a) Models of analyzed pellets; (b) distinguishing of shape parameters of analyzed model of pellet.

The general volume of the pellet is calculated by the previously used method of the sum of the separate volumes of layers, whose height is equal to one point of the image. The improved method is based on the calculation of the ellipse-like areas (based on Figures 8 and 13). The received results are presented in Table 2. The measuring data was compared to the method of the layer sum of circle-like pellets. Data received using the water displacement method is considered benchmark. Measuring results are presented in graphs of Figure 15.

According to the calculations results, for the evaluation of volume of the irregularly shaped pellet, the evaluation method of pellet layer area approximation using the ellipse can be selected as a rational choice. The average absolute error when layers are approximated as circles is 1.191 cm^3 (when the pellet is evaluated rotated by 90° from position, in which the pellet limits the largest possible area in relation to field of view of the camera), which on average represents < 9% of average volume of a pellet. Average squared deviation respectively was 1.38. While, when approximation of layers was carried out using the ellipse geometric shape, the average absolute error was 0.314 cm^3 (when the pellet is evaluated rotated by 90° from position, in which the pellet limits the largest possible area in relation to field of view of the camera) and it represents < 2% of average volume of pellet.

Table 2. Differences between calculation of volume of pellets using direct method and pellet division to circular and elliptical layers.

Sample no.	Sample Position	Volume V1 [2], cm³	Volume V2 [3], cm³	|V1−V2|, cm³	Volume V3 [4], cm³	|V1−V3|, cm³	Volume V4 [5], cm³	|V1−V4|, cm³	Volume V5 [6], cm³	|V1−V5|, cm³
1	Largest area [1]	12.8	14.60	1.80	14.27	1.47	20.09	7.29	19.02	6.22
	Rotated by 90°						10.2	2.6	11.29	1.51
2	Largest area [1]	13.6	14.94	1.34	14.69	1.09	19.56	5.96	19.3	5.7
	Rotated by 90°						11.89	1.71	13.06	0.54
3	Largest area [1]	12.4	13.67	1.27	13.40	1.00	18.07	5.67	17.87	5.47
	Rotated by 90°						10.61	1.79	12.29	0.11
4	Largest area [1]	15.8	14.70	1.10	14.60	1.20	15.45	0.35	14.51	1.29
	Rotated by 90°						15.5	0.3	16.18	0.38
5	Largest area [1]	13.4	13.32	0.08	13.23	0.17	15.14	1.74	14.94	1.54
	Rotated by 90°						12.44	0.96	13.61	0.21
6	Largest area [1]	14.6	15.00	0.40	14.87	0.27	17.46	2.86	17.06	2.46
	Rotated by 90°						13.78	0.82	14.79	0.19
7	Largest area [1]	13.2	13.27	0.07	13.10	0.10	15.79	2.59	15.61	2.41
	Rotated by 90°						11.73	1.47	13.11	0.09
8	Largest area [1]	14.0	14.27	0.27	14.14	0.14	16.65	2.65	16.46	2.46
	Rotated by 90°						12.83	1.17	13.99	0.01
9	Largest area [1]	14.4	13.70	0.70	13.70	0.70	13.79	0.61	13.63	0.77
	Rotated by 90°						14.28	0.12	14.4	0
10	Largest area [1]	14.6	14.28	0.32	14.20	0.40	16.04	1.44	15.82	1.22
	Rotated by 90°						13.63	0.97	14.7	0.1
	Largest area	MAE		0.735		0.654		3.116		2.954
		RMSE		0.932		0.811		3.851		3.537
	Rotated by 90°	MAE		0.735		0.654		1.191		0.314
		RMSE		0.932		0.811		1.380		0.532

[1] Pellet model rotated to limit the largest area in relation to camera. [2] Volume of pellet model calculated using water displacement method. [3] Cross-section layers of pellet models approximated using circles, while diagonals are evaluated according to information of different images. [4] Cross-section layers of pellet models approximated using ellipses, while diagonals are evaluated according to information of different images. [5] Pellet model cross-section layers approximated using circles. [6] Pellet model cross-section layers approximated using ellipses.

To evaluate the practical value of this method, measurements in a real production line were carried out. Measuring equipment, designed, manufactured and installed in previous works, with linear video camera, LED light and automatic subsystem of sampling was used [34]. Monoammonium phosphate samples (MAP) were analyzed during the experiment. Received results were compared to results received using the sieves method and the method of processing of scanned images (where pellets were approximated using ellipse) (Figures 16–18). Same samples were compared during measuring.

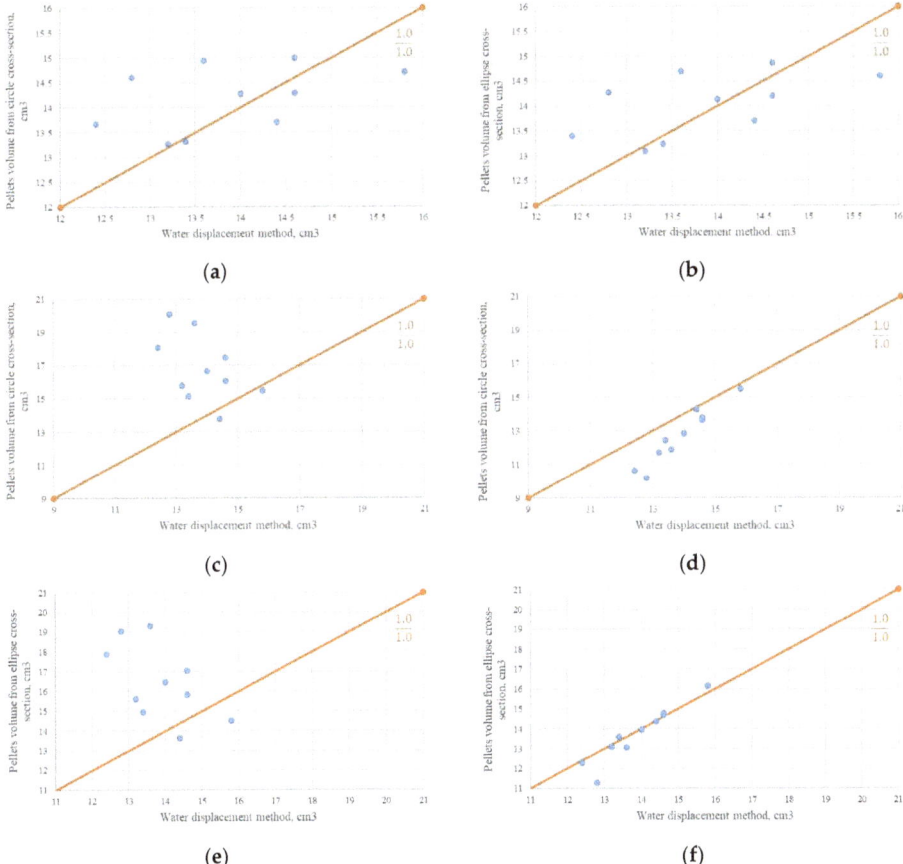

Figure 15. Measuring of volume of formed pellet models based on method of image processing. All graphs present comparison of results to results received using water displacement volume evaluation method: (**a**) cross-sections of pellet model layers approximated using circles; (**b**) cross-sections of pellet model layers approximated using ellipses. In the (**a**,**b**) graphs, particle volume calculation was made by evaluating 2 images (largest area and rotated by 90°). In (**c**,**d**)—cross-sections of pellet model layers approximated using circles from 1 image (largest area or rotated by 90°). In (**e**,**f**)—cross-sections of pellet model layers approximated using ellipses from 1 image (largest area or rotated by 90°).

The evaluation of the received results showed evident matching of the results with those received from the control equipment. Evaluation of the results of separate samples showed that absolute error does not exceed 0.15 mm of the average size of the pellet (D_{50}). The relative error is less than 5%, when the absolute error of approximation of the pellet using the ellipse is up to 0.37 and relative error is > 12%. Therefore, it can be stated that the proposed method for evaluating separate pellets allows for achieving better matching between the measuring results and control equipment results.

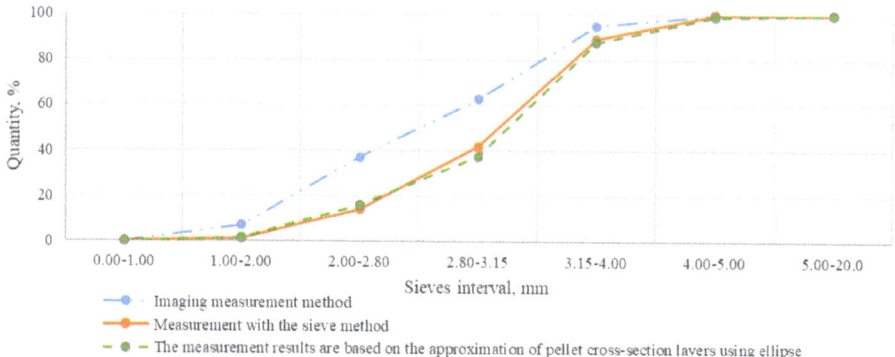

Figure 16. Distribution of pellets according to volume.

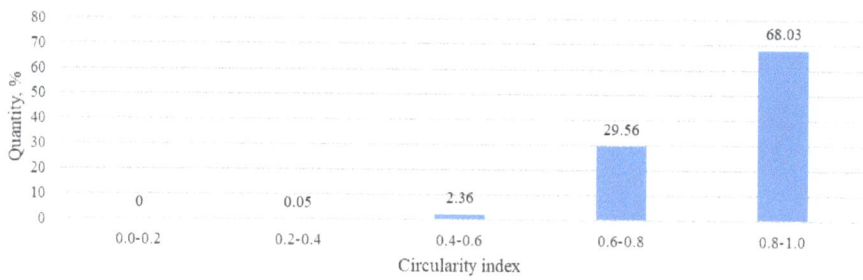

Figure 17. Distribution of pellets according to roundness.

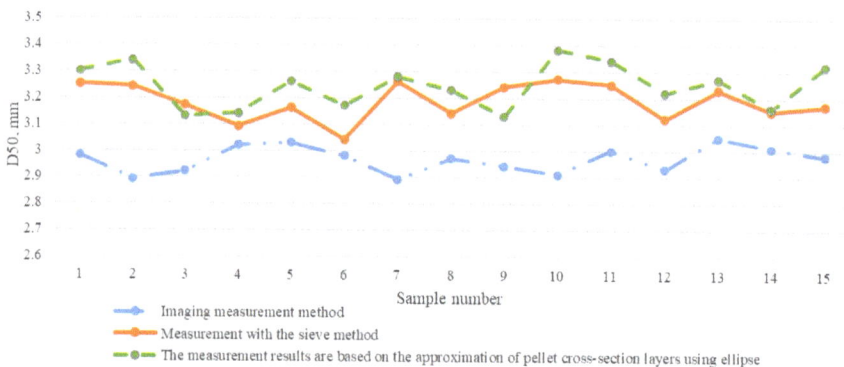

Figure 18. Average diameter of pellet from different samples (D_{50}).

4. Discussion

The received examination results showed that not only the shape of approximation has a great effect on the precision of evaluation, but also the position of the pellet in relation to the video camera. When a less characteristic image of a pellet is scanned, it does not provide sufficient valuable geometric information which could be interpreted correctly. After evaluating measurement results, it was noted that a similar level of discrepancy is found with both approximation of layers using circle and approximation using ellipse. The average absolute error when layers are evaluated as circles was 3.116 cm³, while the relative error is up to 17.40%. When layers are evaluated using the ellipse shape average the absolute error is 2.954 cm³, while the relative error is up to 17.20%. Therefore, the assumption is made that the area and volume of a pellet at a place of notch should be evaluated as a geometric shape of ellipse, because it provides more precise results of measurements. These results can later be used for management of other, more complex processes [41]. The proposed more accurate method of evaluating irregularly shaped pellets can be ensured by implementing a computer vision quality monitoring system in the fertilizer production line. Monitoring the results allows operators to adjust the line parameters that determine the shape of the particles. Ensuring the quality control of production processes determines the quality of the final product. High-quality fertilizer production influences better absorption of useful substances for plants, and more even fertilization.

5. Conclusions

We have provided the results of experiments with the proposed method for evaluation of pellet volume using two-dimensional image processing. After completion of the first experiments it was determined that approximation of irregularly shaped pellets using circular geometric shapes presents a larger amount of error than approximation using ellipsoids. Respectively, when sphere shape approximation is used, the average absolute error is up to 1.258 cm^3 (σ = 0.687), while when using ellipsoid, 0.153 cm^3 (σ = 0.171). It is more than eight times the difference. Additionally, it must be noted that irregularly shaped pellets were evaluated.

The standard method, which is not precise when irregular shapes are being evaluated, was evaluated according to literature sources. Evaluation of pellet volume by using its cross-section layers whose height is limited by the lowest possible height of scanning equipment provides better results of volume calculation. It was determined that the shape of an elongated ellipse is characteristic to a monoammonium phosphate pellet. Additionally, the most common met irregularities of pellet shape (notches) were evaluated. The proposed method allows for the elimination of discrepancies by evaluating the envelope line of the region limited by the pellet silhouette. After improving approximation of the cross-section using the ellipse geometric shape according to convexity of area limited by the pellet's silhouette, better results were achieved. This allowed us to reduce relative error from 10.20% to 2.50%. However, in order to achieve these results, it is very important that the more characteristic region of the scanned pellet should be evaluated. When the pellet is rotated by 90° from position in which this pellet limits the largest area in relation to the field of view of the video camera, the error is only up to < 2% of its average volume. The proposed method of pellets analysis makes it possible to achieve high results of indirect measurements with an error of less than 2%. This provides an opportunity for continuous monitoring of production. To ensure reliable results, they can be compared with the data of measurements performed by the direct measurement method (the latter are performed significantly less frequently). In further works it is planned to carry out experiments with pellet ducts of different shapes, which are used to transport pellet for testing.

Author Contributions: Conceptualization, D.A. and A.L.; methodology, A.L. and D.A.; software, A.L. and D.N.; validation, A.L., M.Z., and V.M.; formal analysis, R.S. and J.J.; visualization, M.Z., D.N., and D.K.; investigation, A.V., V.M., J.J., and D.K.; resources, A.L. and A.V.; data curation, A.L.; writing—original draft preparation, D.A. and A.L.; writing—review and editing, V.M., D.K. and R.S.; supervision, D.A.; funding acquisition, A.V., M.Z., and D.N. All authors have read and agreed to the published version of the manuscript.

Funding: This research received no external funding.

Conflicts of Interest: The authors declare no conflict of interest.

References

1. Markovic, S.; Stojanovic, Z.S. Determination of Particle Size Distributions by Laser Diffraction. *Tech. New Mater.* **2012**, *21*, 11–20.
2. Witt, W.; Heuer, M.; Schaller, M. In-line particle sizing for process control in new dimensions. *China Particuol.* **2004**, *2*, 185–188. [CrossRef]
3. Cornillault, J. Particle Size Analyzer. *Appl. Opt.* **1972**, *11*, 265. [CrossRef] [PubMed]

4. Coghill, P.; Millen, M.; Sowerby, B. On-line measurement of particle size in mineral slurries. *Miner. Eng.* **2002**, *15*, 83–90. [CrossRef]
5. McClements, D.J. Ultrasonic Measurements in Particle Size Analysis. In *Encyclopedia of Analytical Chemistry: Applications, Theory and Instrumentation*; John Wiley & Sons, Ltd.: Hoboken, NJ, USA, 2006.
6. Wan, Q.; Jiang, W. Near field acoustic holography (NAH) theory for cyclostationary sound field and its application. *J. Sound Vib.* **2006**, *290*, 956–967. [CrossRef]
7. Findlay, W.P.; Peck, G.R.; Morris, K.R. Determination of fluidized bed granulation end point using near-infrared spectroscopy and phenomenological analysis. *J. Pharm. Sci.* **2005**, *94*, 604–612. [CrossRef]
8. Petrak, D. Simultaneous Measurements of Particle Size and Velocity with Spatial Filtering Technique In Comparison With Coulter Multisizer and Laser Doppler Velocimetry. In Proceedings of the 4th International Conference on Multiphase Flow, New Orleans, LA, USA, 27 May–1 June 2001. Available online: https://www.parsum.de/wp-content/uploads/2017/10/ConfMPF882.pdf (accessed on 1 March 2020).
9. Dieter, P.; Stefan, D.; Günter, E.; Michael, K. In-line particle sizing for real-time process control by fibre-optical spatial filtering technique (SFT). *Adv. Powder Technol.* **2011**, *22*, 203–208. [CrossRef]
10. Bishop, D. A sedimentation method for the determination of the particle size of finely divided materials (such as hydrated lime). *Bur. Stand. J. Res.* **1934**, *12*, 173. [CrossRef]
11. Maab, S.; Rojahn, J.; Emmerich, J.; Kraume, M. On-Line Monitoring Of Fluid Particle Size Distributions In Agitated Vessels Using Automated Image Analysis. In Proceedings of the 14th European Conference on Mixing, Warszawa, Poland, 10–13 September 2012. Available online: https://pdfs.semanticscholar.org/dce5/c4ca8f3f676f6a65132654ff59351b11ee88.pdf (accessed on 1 March 2020).
12. Lu, Z.; Hu, X.; Lu, Y. Particle Morphology Analysis of Biomass Material Based on Improved Image Processing Method. *Int. J. Anal. Chem.* **2017**, *2017*, 1–9. [CrossRef]
13. Hijazi, B.; Cool, S.; Vangeyte, J.; Mertens, K.; Cointault, F.; Paindavoine, M.; Pieters, J. High Speed Stereovision Setup for Position and Motion Estimation of Fertilizer Particles Leaving a Centrifugal Spreader. *Sensors* **2014**, *14*, 21466–21482. [CrossRef]
14. Närvänen, T.; Seppälä, K.; Antikainen, O.; Yliruusi, J. A New Rapid On-Line Imaging Method to Determine Particle Size Distribution of Granules. *AAPS PharmSciTech* **2008**, *9*, 282–287. [CrossRef] [PubMed]
15. Laitinen, N.; Antikainen, O.; Yliruusi, J. Characterization of particle sizes in bulk pharmaceutical solids using digital image information. *AAPS PharmSciTech* **2003**, *4*, 383–391. [CrossRef] [PubMed]
16. Watano, S.; Numa, T.; Miyanami, K.; Osako, Y. On-line Monitoring of Granule Growth in High Shear Granulation by an Image Processing System. *Chem. Pharm. Bull.* **2000**, *48*, 1154–1159. [CrossRef] [PubMed]
17. Kumari, R.; Rana, N. Particle Size and Shape Analysis using Imagej with Customized Tools for Segmentation of Particles. *Int. J. Eng. Res.* **2015**, *4*, 247–250.
18. Mazzoli, A.; Favoni, O. Particle size, size distribution and morphological evaluation of airborne dust particles of diverse woods by Scanning Electron Microscopy and image processing program. *Powder Technol.* **2012**, *225*, 65–71. [CrossRef]
19. Habrat, M.; Mlynarczuk, M. Object Retrieval in Microscopic Images of Rocks Using the Query by Sketch Method. *Appl. Sci.* **2019**, *10*, 278. [CrossRef]
20. Okarma, K. Current Trends and Advances in Image Quality Assessment. *Elektronika ir Elektrotechnika* **2019**, *25*, 77–84. [CrossRef]
21. Laucka, A.; Adaskeviciute, V.; Valinevicius, A.; Andriukaitis, D. Research of the Equipment Calibration Methods for Fertilizers Particles Distribution by Size Using Image Processing Measurement Method. In Proceedings of the 2018 23rd International Conference on Methods & Models in Automation & Robotics (MMAR), Międzyzdroje, Poland, 27–30 August 2018; pp. 407–412.
22. Vrbančič, G.; Podgorelec, V. Automatic Classification of Motor Impairment Neural Disorders from EEG Signals Using Deep Convolutional Neural Networks. *Elektronika ir Elektrotechnika* **2018**, *24*, 3–7. [CrossRef]
23. Forristal, D. Fertiliser Prills or Granules: Which Spread Best? The IFJ Article. 2014. Available online: https://www.teagasc.ie/media/website/publications/2014/Fertiliser_Prills_or_granules.pdf (accessed on 1 March 2020).
24. Particle Size Result Interpretation: Number vs. Volume Distributions. Available online: https://www.horiba.com/scientific/products/particle-characterization/education/general-information/data-interpretation/number-vs-volume-distributions/ (accessed on 1 March 2020).

25. ISO 9276-2:2001: Representation of Results of Particle Size Analysis—Part 2: Calculation of Average Particle Sizes/Diameters and Moments from Particle Size Distributions. Available online: https://www.iso.org/standard/33997.html (accessed on 1 March 2020).
26. ASTM E 799-03 Standard Practice for Determining Data Criteria and Processing for Liquid Drop Size Analysis. Available online: https://www.astm.org/Standards/E799.htm (accessed on 1 March 2020).
27. International Standard ISO 20998-1:2006 "Measurement and Characterization of Particles by Acoustic Methods". Available online: https://www.iso.org/standard/39869.html (accessed on 1 March 2020).
28. Stewart, L.; Bandel, V.A. Uniform Lime and Fertilizer Spreading. Publication about Fertilizers Equipment and Production Process. Available online: https://extension.umd.edu/sites/extension.umd.edu/files/_images/programs/anmp/EB_254_Uniform%20Lime%20and%20Fertilizer%20Spreading.pdf (accessed on 1 March 2020).
29. Das, S.C.; Behara, S.R.B.; Morton, D.A.V.; Larson, I.; Stewart, P. Importance of particle size and shape on the tensile strength distribution and de-agglomeration of cohesive powders. *Powder Technol.* **2013**, *249*, 297–303. [CrossRef]
30. Yu, H.; Ding, Y.; Liu, Z.; Fu, X.; Dou, X.; Yang, C. Development and Evaluation of a Calibrating System for the Application Rate Control of a Seed-Fertilizer Drill Machine with Fluted Rollers. *Appl. Sci.* **2019**, *9*, 5434. [CrossRef]
31. Jannat, E.; Arif, A.A.; Hasan, M.M.; Zarziz, A.B.; Rashid, H.A. Granulation techniques & its updated modules. *Pharma Innov. J.* **2016**, *5*, 134–141. Available online: https://pdfs.semanticscholar.org/9595/861fa792566927030aabdc9d0cb63a55ce59.pdf (accessed on 1 March 2020).
32. Le, T.-T.; Miclet, D.; Héritier, P.; Piron, E.; Chateauneuf, A.; Berducat, M. Morphology characterization of irregular particles using image analysis. Application to solid inorganic fertilizers. *Comput. Electron. Agric.* **2018**, *147*, 146–157. [CrossRef]
33. Igathinathane, C.; Pordesimo, L.; Columbus, E.; Batchelor, W.; Methuku, S.; Cannayen, I. Shape identification and particles size distribution from basic shape parameters using ImageJ. *Comput. Electron. Agric.* **2008**, *63*, 168–182. [CrossRef]
34. Laucka, A.; Adaskeviciute, V.; Andriukaitis, D. Research of the Equipment Self-Calibration Methods for Different Shape Fertilizers Particles Distribution by Size Using Image Processing Measurement Method. *Symmetry* **2019**, *11*, 838. [CrossRef]
35. Hogg, R. A Spheroid Model for the Role of Shape in Particle Size Analysis. *KONA Powder Part. J.* **2015**, *32*, 227–235. [CrossRef]
36. Riddle, D.F. *Calculus and Analytic Geometry*; Wadsworth Publishing Company, Inc.: Belmont, CA, USA, 1979; p. 505. ISBN 9780534006266.
37. Rashidi, M.; Gholami, M. Determination of kiwifruit volume using ellipsoid approximation and Image-processing methods. *Int. J. Agric. Biol.* **2008**, *10*, 375–380. Available online: https://www.researchgate.net/publication/228409809_Determination_of_kiwifruit_volume_using_ellipsoid_approximation_and_image-processing_methods (accessed on 2 March 2020).
38. Sabliov, C.M.; Boldor, D.; Keener, K.M.; Farkas, B.E. Image processing method to determine surface area and volume of axi-symmetric agricultural products. *Int. J. Food Prop.* **2002**, *5*, 641–653. [CrossRef]
39. Lech, P.; Okarma, K. Prediction of the Optical Character Recognition Accuracy based on the Combined Assessment of Image Binarization Results. *Elektronika ir Elektrotechnika* **2015**, *21*, 62–65. [CrossRef]
40. Khitas, M.; Ziet, L.; Bouguezel, S. Improved Degraded Document Image Binarization Using Median Filter for Background Estimation. *Elektronika ir Elektrotechnika* **2018**, *24*, 82–87. [CrossRef]
41. Andriukaitis, D.; Laucka, A.; Valinevicius, A.; Zilys, M.; Markevicius, V.; Navikas, D.; Sotner, R.; Petrzela, J.; Jerabek, J.; Herencsar, N.; et al. Research of the Operator's Advisory System Based on Fuzzy Logic for Pelletizing Equipment. *Symmetry* **2019**, *11*, 1396. [CrossRef]

© 2020 by the authors. Licensee MDPI, Basel, Switzerland. This article is an open access article distributed under the terms and conditions of the Creative Commons Attribution (CC BY) license (http://creativecommons.org/licenses/by/4.0/).

Article

Histogram-Based Descriptor Subset Selection for Visual Recognition of Industrial Parts

Ibon Merino [1,2,*], Jon Azpiazu [1], Anthony Remazeilles [1] and Basilio Sierra [2]

1. TECNALIA, Basque Research and Technology Alliance (BRTA), Paseo Mikeletegi 7, 20009 Donostia-San Sebastian, Spain; jon.azpiazu@tecnalia.com (J.A.); anthony.remazeilles@tecnalia.com (A.R.)
2. Department of Computer Science and Artificial Intelligence, University of the Basque Country UPV/EHU, 20018 Donostia-San Sebastian, Spain; b.sierra@ehu.eus
* Correspondence: ibon.merino@tecnalia.com

Received: 1 April 2020; Accepted: 25 May 2020; Published: 27 May 2020

Abstract: This article deals with the 2D image-based recognition of industrial parts. Methods based on histograms are well known and widely used, but it is hard to find the best combination of histograms, most distinctive for instance, for each situation and without a high user expertise. We proposed a descriptor subset selection technique that automatically selects the most appropriate descriptor combination, and that outperforms approach involving single descriptors. We have considered both backward and forward mechanisms. Furthermore, to recognize the industrial parts a supervised classification is used with the global descriptors as predictors. Several class approaches are compared. Given our application, the best results are obtained with the Support Vector Machine with a combination of descriptors increasing the F1 by 0.031 with respect to the best descriptor alone.

Keywords: computer vision; feature descriptor; histogram; feature subset selection; industrial objects

1. Introduction

Computer vision, in the last years, has gained much interest in many fields, such as autonomous driving [1], medical [2], face recognition [3], object detection [4], and object segmentation [5]. Perception is also regarded as one of the key enabling technologies for extending the robot capabilities, preferentially targeting flexibility, adaptation, and robustness, as required for fulfilling the industry 4.0 paradigm [6]. Although in most fields large and complex datasets can be obtained, detection of industrial parts has a lack of datasets. One of the reasons is that most of the time in industrial context, the aim is to detect an object from which usually the CAD is available. However, sometimes there is a need of detecting diverse, complex, and tiny objects [7] and lack of time to generate a robust dataset (taking pictures and labeling). One of the solutions is to generate simulated data to train the models but usually there is a significant gap transferring that learned knowledge to reality.

To make matter worse, industrial parts are usually texture-less. This means that many of the most used recognition methods cannot deal with them. One of the methods to deal with texture-less objects are Convolutional Neural Networks. Nowadays, computer vision researches are mainly focused on using Convolutional Neural Networks (CNN) [8–10]. One of the disadvantages of the CNNs is the need of a large dataset to train them. Even if it is possible to use the CNN trained on other fields in industry [11], there is still a need of a large enough training dataset to obtain good results. Feature descriptors based on classical methods have been very useful and thoroughly spread in the literature previous to CNN. One of the benefits of using this approach is that there is no need of a large training set to obtain good results. Actually, there are many image descriptors and each of them has its advantages and disadvantages.

Our approach is based in the idea that the combination of different descriptors leads to a better performance, taking advantage of the benefits of each descriptor to deal with the two problems mentioned before (lack of a large dataset and texture-less objects). The crux of the matter is to select the descriptors that contribute to achieve a better result and discard those that do not provide any improvement. Our method achieves a classification quality similar to state-of-the-art methods on the experiments done.

In Section 2, we present a background of the description methods, classifiers, and features subset selection techniques. In Section 3, we explain the combination of the descriptors and the image classification. The experiments done and their results are gathered in Section 4. Finally, in Section 5, the conclusions are summarized.

2. Background

The analysis of images usually relies on the extraction of visual features. Such an approach can be observed in classification [12], object detection [4], and segmentation [5]. In this section, we provide an overview of the main feature descriptors, together with some of the related classification techniques.

2.1. Features Descriptors

Local features extractors are characteristic local primitives as points focusing on a close neighborhood. Some examples of those features are SIFT [13], SURF [14], and LBP [15]. Global descriptors, instead, extract information directly from the whole image by computing histograms for example. Local features are good for image recognition as each point is independent from the rest and the features are more discriminant. Global features instead are more used for classification and object detection as they achieve a more global representation. Nevertheless, small changes have a larger impact on global features and a better preprocessing is needed when using them. Extracting global features and their classification is usually faster.

As a matter of a fact, combining both local and global features usually performs better [16]. Many researchers use histograms of local features to obtain benefits of both types. Doing so, we obtain a global representation of the local features. [16] present a taxonomy called Histogram of Equivalent Patterns (HEP) that gathers those histograms of local features. In order for a feature to be part of this framework, it needs to have a delimited quantification, that is, the number of possible values of the extracted feature must be small enough to obtain a relevant histogram. For example, LBP [15] is part of this framework as the possible values are 256 so the resulting histogram is of length 256, while HOG or SIFT are not part of the HEP framework as the number of possible values is high and the resulting histogram is not relevant. In [17], a combination of descriptors was also used, but limited to local descriptors.

One of the first HEP methods was introduced in 1973. This method, called Gray Level Co-occurrences Matrices (GLCM) [18] measures the joint probability of the gray levels of two pixels standing in some predefined relative positions. Since 1973, it has been widely used in many texture analysis applications as a feature extractor in this context.

In 1990, [19] proposed the texture spectrum (TS), which inspired many HEP methods. This texture descriptor is based in decomposing the image into a set of essential small units, called Texture Units (TUs). The occurrence distribution of TU is the TS. One of the first and most used TU-based descriptors is the Local Binary Pattern (LBP) [15]. This last one is a two-level TU, gray-scale invariant and easily combined with a simple contrast measure. One of the main characteristics is its robust invariant to light changes.

Another method based in the TU is the Simplified Texture Unit (STU) [20]. This method use a more reduced range of values without a significant loss of the characterization power. This way, there are two options of STU: using the crosswide neighbors (up, right, down, and left) and using diagonal neighbors (up-left, up-right, down-right, and down-left); its reduced length is commonly used in real-time applications obtaining similar performance to LBP.

The modified texture spectrum (MTS) [21] can be considered as a simplified version of LBP, where only a subset of the peripheral pixels (up-left, up, up-right, and right) are considered. Its TS is 16 elements in length, significantly improving the computation efficiency on classification. Similarly to STU, the reduction on the TS length leads to a faster classification while achieving similar performance.

The GaborLBP [22] considers the advantages of the Gabor filters in computer vision and exploits them. It first applies a Gabor transformation and encodes the magnitude values with the LBP operator. Fusing both tools enables handling of illumination changes, viewpoint angle changes, and non-rigid bodies. Usually this combination is used for face recognition or person identification.

The Local Ternary Pattern (LTP) [23] is a generalization of the LBP and it is more discriminant and less sensitive to noise in uniform regions. It is a local texture descriptor that uses a 3-value coding that thresholds around zero. Comparing to the LBP, LTP is more resistant to noise but no longer invariant to gray-level transformations.

The Binary Gradient Contours (BGC) [24] is a binary 8-tuple. It relies on computing a set of eight binary gradients between pairs of pixels all along a closed path around the central pixel of a 3×3 grayscale image patch. They defined the closed path in three different ways: single-loop (BGC1), double-loop (BGC2), and triple-loop (BGC3).

Another HEP descriptor, is the Local Quantized Patterns (LQP) [25]. This is a generalization of local pattern features that makes use of vector quantization. It uses large local neighbourhoods and/or deeper quantization with domain-adaptative vector quantization.

The Weber's Law Descriptor (WLD) [26] was proposed in 2010 as a simple, yet very powerful and robust descriptor. It is based on the fact that human pattern perception also depends on the original intensity of the stimulus and not only on the change of a stimulus (such as sound and lighting). It is composed of two components: differential excitation and orientation.

The Histogram of oriented gradients (HOG) [27] is a feature descriptor that counts the occurrences of gradient orientation in localized portions of an image. Operating on local cells provides invariation to geometric and photometric transformations. The HOG descriptor is particularly suited for human detection in images. Even if HOG is not part of HEP, the way it generates the descriptor (calculating a histogram of gradients) works similar to HEP methods so it can be used similarly.

2.2. Classifiers

Descriptors are used to obtain features from images. Those features are then used by the classifiers to predict which object is on each image. Many machine learning algorithms are used for classifying images, but some of the most popular ones are K-Nearest Neighbors, Naive Bayes, Random Forest, Support Vector machine, Random Committee, Bagging, and Multiclass Classifier.

The Nearest Neighbor Rule is a well-known algorithm and the simplest nonparametric decision procedure that assigns to the uncategorized object the label of the closest sample of the training set. In 1967, a modification of this algorithm led to one of the most used classification algorithms, the K-Nearest Neighbors (KNN) [28]. It is based on looking for closest points and classifying them as the majority class. For a given set of n pairs $(x_1, \theta_1), ..., (x_n, \theta_n)$, where x_i is in a metric space X and θ_i is the category that x_i belongs to from a subset $\{1, 2, ..., M\}$, a new arriving instance x is analyzed to estimate its corresponding class θ. This estimation is done by looking for the nearest neighbor $X'_n \in (x_1, x_2, ..., x_n)$:

$$\min d(x_i, x) = d(x'_n, x) \quad i = 1, 2, ..., n$$

where d is a distance metric according to the space X. The new instance x will be assigned to the category θ'_n. This is the basic 1-NN. In general, KNN rule decides x belongs to the category of majority vote of the nearest k neighbors.

The Naive Bayes [29], the simplest Bayesian classifier, is another classification algorithm that is often used for its simplicity. It is based on the Bayesian Rule and assumes that variables are independent given the class. Despite this unrealistic assumption, it is successful in practice. The Bayesian rule states that the probability that a instance x belongs to class C_k is

$$P(C_k|x) = \frac{P(C_k)P(x|C_k)}{P(x)} \quad (1)$$

where C_k is the class between the K possible classes and x the instance to be classified. Taking into account the independence assumption, the conditional distribution over the class variable C is

$$p(C_k|x_1, ..., x_n) = \frac{1}{Z} p(C_k) \prod_{i=1}^{n} p(x_i|C_k) \quad (2)$$

The instance is classified as the class with more $p(C_k|x_1, ..., x_n)$.

The Random Forest (RF) [30] is a combination of decision trees that use random subsets of the features to be built. Figure 1 shows an example of RF.

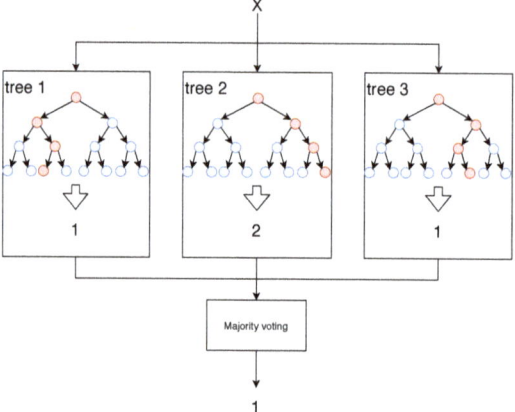

Figure 1. Random Forest example where each tree classifies the new instance and the resulting class is decided by majority voting.

Support Vector Machines (SVM) [31] are supervised learning models that look for optimal hyperplanes that separates classes. An optimal hyperplane is defined as the linear decision function with maximal margin between the vectors of the two classes (Figure 2).

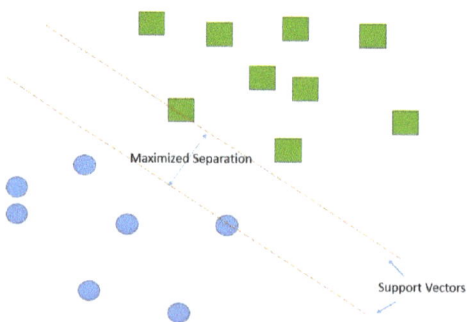

Figure 2. Support vector machine: maximum separation between two classes.

Random Committee (RC) [32] is a committee of random classifiers. The base randomizable classifiers (that form the committee members) are built using different random number seeds based in the same data. The final prediction is a straight average of the predictions generated by the individual base classifiers.

The Bagging [33] technique is called after Bootstrap aggregating. This machine learning ensemble that can be used to improve the stability of a model by improving the accuracy and reducing variance in order to reduce overfitting.

2.3. Feature Selection

As stated before, the crux of the matter in this paper relies on how to select the different visual features to improve the individual score of each descriptor. Some authors have used different techniques to do this [34,35]. Feature Selection is a machine learning technique that is used in many fields and usually improves the accuracy of the model. In [34], the authors uses different feature selection techniques to improve the score in the Quantitative Structure–Activity Relationship (QSAR). In [35], instead, they use a similar approach for hand pose recognition. In [36], a view over the different feature selection techniques and its variations is described. Our approach is based in those methods and is used in a completely different context.

3. Proposed Approach

In order to achieve a better performance than just using a single global descriptor, we propose using a Descriptor Subset Selector. That is, we try to find the combination of global descriptors that scores a better result. Among all available options of subset selection, we have used 2 for their greedy approach which achieve a significant performance: forward selection and backward selection. First, we present the classification of a single image, given a descriptor and a classifier. After that, we explain the feature selection techniques to choose the combination of descriptor to use. Next, we present the evaluation methods, in order to decide which is the best solution. Finally, we present the whole pipeline of the proposed approach.

3.1. Classification

The first step in the pipeline is to classify a picture into the C different classes. Given a descriptor and a classifier, the classifier is trained with features obtained from the description of the set of images for training. Given a new image to be classified, the descriptor extracts the feature from the image and that feature is classified by the classifier (Figure 3).

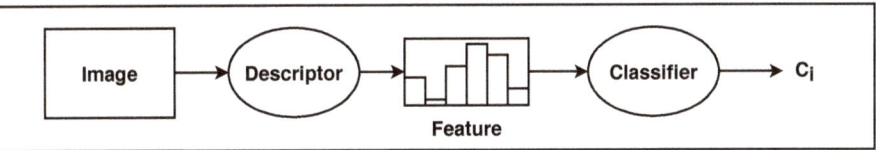

Figure 3. Classification of a new image given a descriptor and a classifier.

3.2. Feature Selection Techniques

The feature selection techniques are used to chose the descriptors for the classification. An exhaustive search of best combination of descriptors is computationally inefficient, while it guarantees that the optimal solution is achieved. Nevertheless, a suboptimal solution can be achieved using a sequential search. This is an iterative search that once a stage of the search is reached, is impossible to go back. The complexity of the exhaustive search is exponential ($O(2^n)$), while the sequential search remains polynomial ($O(n^{k+1})$), where k is the number of evaluated subsets in each stage. This last one does not guarantee an optimal solution.

Another important consideration in the feature selection techniques is the generation of the successors, i.e., how to select the next candidates for the following stage. The simplest and most used methods are Forward and Backward generation [36]. In forward generation, on each stage the element which makes J (the evaluation measure) greater is selected and added to the selected subset. For example, the first descriptor added to the subset would be the one with the best individual score. The next stage would add to the subset the one that concatenated with the previous one makes the score greater. We refer to this method as Sequential Forward Subset Selection (SFSS) [36], and its pseudocode is described in Algorithm 1. The backwards is the opposite behavior. The subset is initialized with all the elements and on each stage the element that that makes J greater when removed is done so. The stopping criteria in both cases can be that J is not increased in j steps or the subset achieves a desired length. We refer to this method as Sequential Backward Subset Selection (SBSS) [36], and its pseudocode is described in Algorithm 2.

Algorithm 1: Sequential Forward Subset Selection

Input :
    X—Set of elements
    J—evaluation metric

Output:
    X′—solution found

$X' = \emptyset$

repeat
    $x' := argmax\{J(X' \cup x) | x \in (X \setminus X')\}$
    $X' := X' \cup \{x'\}$
until *not improvement in J OR* $X' = X$;

---

where $\cup$ stands for union between two sets or an element and a set and $\setminus$ operator stands for difference.

Algorithm 2: Sequential Backward Subset Selection

Input :
    X—Set of elements
    J—evaluation metric

Output:
    X′—solution found

$X' = X$

repeat
    $x' := argmax\{J(X' \setminus x) | x \in X'\}$
    $X' := X' \setminus \{x'\}$
until *not improvement in J OR* $X' = \emptyset$;

3.3. Evaluation Measure

A classification quality can be quantified using measures such the one of Equation (3). This measure, named F-value [37] or F-score, is an evaluation measure that takes into account the precision and the recall. More precisely, the metric used is a particular case of the F-value where the precision and the recall are balanced. This is called F_1, an harmonic mean between the precision and the recall.

$$F_1(y) = 2 \cdot \frac{precision_y * recall_y}{precision_y + recall_y} \qquad (3)$$

where y refers to a class (also referred in this paper as C_i). F_1 is class-dependent, so for each class, y, the precision and the recall are computed for that class. The precision (Equation (4)) is the ratio between the correctly predicted views with label y (tp_y or true positive) and all predicted views for that given instance ($|\psi(X) = y|$). The recall (Equation (5)), instead, is the relation between correctly predicted views with label y (tp_y or true positive) and all views that should have that label ($|label(X) = y|$).

$$precision_y = \frac{tp_y}{|\psi(X) = y|} \quad (4)$$

$$recall_y = \frac{tp_y}{|label(X) = y|} \quad (5)$$

To evaluate each stage of the feature selection we use the averaged F_1. This is the mean of the F_1's of all the classes (Equation (6)).

$$F_1 = \frac{1}{|Y|} \sum_{y \in Y} F_1(y) \quad (6)$$

3.4. Full Pipeline

The dataset is divided in two sets: training and test. During the search of the best combination of descriptors, training set is used for training the classifiers and validate the feature selection technique. This separation is made by a Leave-One-Out Cross-Validation (LOOCV) [38]. Each image of the set is used as validation while the rest of the set is used to train the model. Figure 4 shows the whole process. Given a descriptor and a classifier, both are tested using the LOOCV to set the training and validation sets. Once the best combination of descriptors is found, to test the quality of this combination, we use the test set to obtain a general evaluation metric.

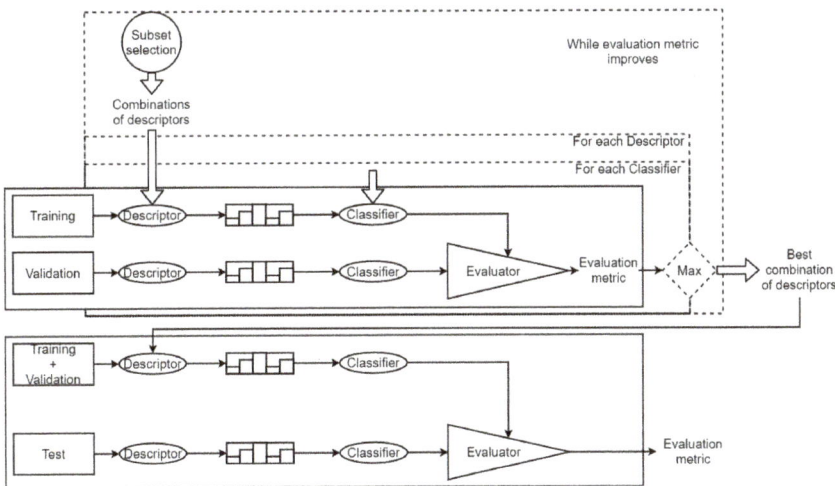

Figure 4. Full pipeline of the proposed method, including training, validation, and evaluation.

4. Experiments and Results

As stated before, the aim of this paper is to present a method to improve the accuracy on reduced datasets of texture-less objects. In order to prove that our method improves the score of the descriptors by their own, we have created a small dataset composed by seven different random industrial parts (Figure 5). We took 50 pictures of each industrial part taken from different viewpoints and different illumination conditions. Objects are rotated and translated but all images are free from occlusion, and with an empty and white background.

Figure 5. Pictures of the parts used in the experiment.

Our pool of descriptors D for discovering the best combination is made up of BGC1 BGC2, BGC3, LBP, GaborLBP, GLCM, HOG, LQP, LTP, MTS, STU+ (or STU1), STU $\times$ (or STU2), and WLD. All descriptors but HOG are computed on grids of different sizes: 1×1, 4×4, and 8×8. The length of gridded histograms is the length of the descriptor multiplied by the number of grids. The HOG is applied to the whole image directly. Figure 6 shows a sample image from our database that has been described by each of the descriptors.

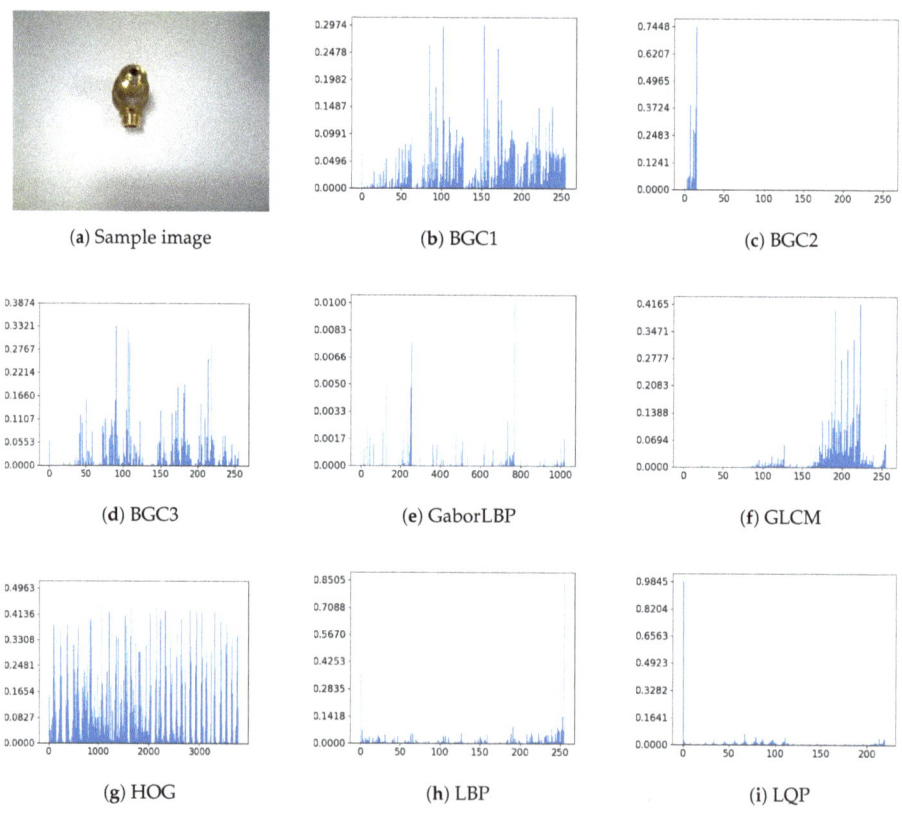

Figure 6. *Cont.*

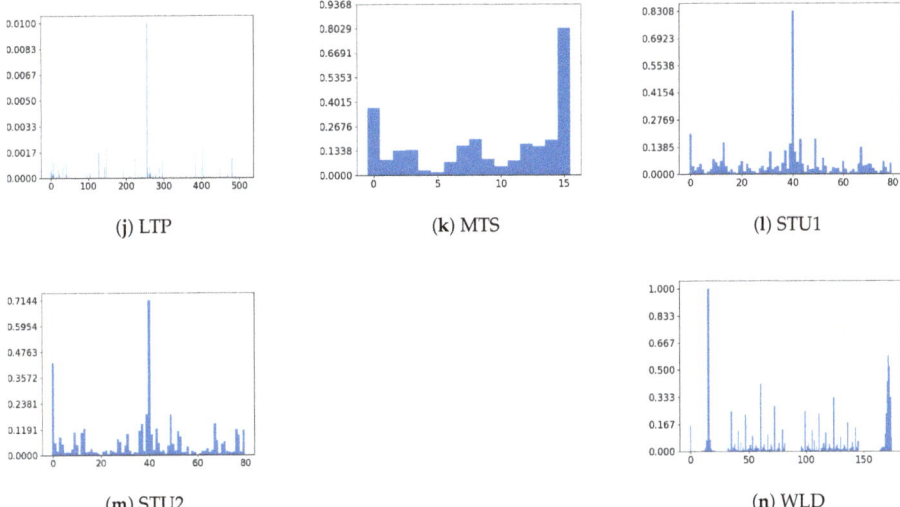

Figure 6. Histogram of all the used descriptors applied to a sample image. The vertical axis represents the number of occurrences of each texture unit normalized and the horizontal axis represents each of the texture units of the histograms. The descriptors are the ones that are part of D described at the beginning of Section 4.

The classifiers used are KNN, NB, SVM 1-vs-1 trained with SMO (Sequential Minimal Optimization [39]), SVM 1-againt-all trained with SGD (Stochastic Gradient Descent [40]), RC, RF, and Bagging. To distinguish between the two SVM implementations, we call SVM to the one trained with SMO and SVM-SGD to the other one. In terms of performance, some of the classifiers are drastically affected by the parameters, but tuning the parameters makes a complex casuistry which is not the aim of this paper. Used parameters are standards and those are given in the Appendix A. The results are obtained for a Intel Xeon CPU of 3GHz and 16GB of RAM, and no GPU acceleration has been used. The following subsections explain the results obtained in the experiments.

4.1. Forward Subset Selection

Forwards Subset Selection of descriptors applied to the whole image (from now on, $FSS1 \times 1$) experiments results are shown in Table 1. In Table 1, the classifier that is between brackets is the one that achieves the highest mean score. If we would use the best descriptor alone, the F_1 would be 0.94 with WLD. By combining it with BGC2 and MTS, and using SVM as classifier, we are able to augment quality of 3% to reach 0.971. On first iteration WLD outperforms the other descriptors with a difference of 0.1 comparing to the next best descriptor. The second iteration increases the overall accuracy and in almost all the cases improves the accuracy of the previous iteration best case.

Table 2 shows the results of the Forwards Subset Selection of descriptors applied to a 4×4 grid ($FSS4 \times 4$). On average, the first iteration performs better than the non-gridded version $FSS4 \times 4$, but the last iteration does not improve the results obtained with $FSS4 \times 4$. The first iteration achieves an F_1 of 0.934 and the final iteration 0.969. Therefore, an improvement of 3.5% is obtained. The final combination of descriptors, the one which achieves the highest score, is composed by STU1 and WLD.

Table 3 shows the results for the 8×8 gridded version ($FSS8 \times 8$). The results are similar to the ones obtained in $FSS4 \times 4$. The first iteration achieves an F_1 of 0.94, while the last one achieves a score of 0.96. In this case, the improvement is 2%.

The performance of the 3 options of the parameters are similar but the speed of the classification is much faster with the $FSS1 \times 1$ version because the length of the final descriptor is shorter. Therefore, the value of the grid parameter makes not a significant difference in the performance. The recommendation is to use the *FSS*1 × 1.

Table 1. Forward Subset Selection of descriptors applied to the whole image (also known as *FSS*4 × 4). Level 1 uses only one descriptor. The following levels concatenate the best descriptor from the previous level to the rest of the descriptors. The algorithm stops on level 4 because the evaluation measure is not improved from level 3 to 4.

Descriptor	Level 1	Level 2 WLD +	Level 3 WLD + BGC2 +	Level 4 WLD + BGC2 + MTS +
BGC1	0.66 (RF)	0.931 (RF)	0.937 (SVM)	0.934 (SVM/RF)
BGC2	0.489 (SVM)	**0.969 (SVM)**	—	—
BGC3	0.611 (SVM)	0.937 (RC)	0.929 (RF)	0.929 (RF)
GaborLBP	0.671 (RF)	0.931 (RF)	0.903 (RF)	0.906 (RF)
GLCM	0.811 (RF)	0.951 (RF)	0.903 (RF)	0.963 (SVM)
HOG	0.84 (SVM)	0.923 (SVM)	0.923 (SVM)	0.923 (SVM)
LBP	0.611 (RF)	0.946 (RF)	0.917 (SVM)	0.923 (RF)
LQP	0.697 (RF)	0.949 (SVM)	0.954 (SVM)	0.949 (SVM)
LTP	0.563 (RF)	0.966 (SVM)	0.969 (SVM)	**0.966 (SVM)**
MTS	0.666 (KNN)	0.966 (SVM)	**0.971 (SVM)**	—
STU1	0.746 (SVM)	0.951 (SVM)	0.951 (SVM)	0.948 (SVM)
STU2	0.74 (SVM)	0.951 (SVM)	0.957 (SVM)	0.957 (SVM)
WLD	**0.94 (RF)**	—	—	—

Table 2. Forward Subset Selection of descriptors applied to 4 × 4 gridded image (also known as *FSS*4 × 4).

Descriptor	Level 1	Level 2 STU1 +	Level 3 STU1 + WLD +
BGC1	0.877 (SVM)	0.908 (SVM-SGD)	0.934 (SVM)
BGC2	0.903 (SVM)	0.931 (SVM)	**0.969 (SVM)**
BGC3	0.857 (SVM)	0.906 (SVM)	0.931 (SVM)
GaborLBP	0.834 (SVM)	0.883 (SVM)	0.903 (SVM)
GLCM	0.923 (RF)	0.957 (SVM)	0.966 (SVM)
HOG	0.846 (KNN)	0.917 (SVM)	0.94 (SVM)
LBP	0.874 (SVM)	0.906 (SVM)	0.929 (SVM)
LQP	0.911 (SVM)	0.94 (SVM)	**0.969 (SVM)**
LTP	0.889 (SVM)	0.931 (SVM)	0.96 (SVM)
MTS	0.909 (SVM)	0.94 (SVM)	0.96 (SVM)
STU1	**0.934 (SVM)**	—	—
STU2	0.914 (SVM)	0.931 (SVM)	0.96 (SVM)
WLD	0.931 (SVM)	**0.969 (SVM)**	—

Table 3. Forward Subset Selection of descriptors applied to 8 × 8 gridded image (also known as *FSS*8 × 8).

Descriptor	Level 1	Level 2 WLD +	Level 3 WLD + MTS +
BGC1	0.845 (SVM)	0.877 (SVM)	0.897 (SVM)
BGC2	0.911 (SVM)	0.954 (SVM)	0.957 (SVM)
BGC3	0.843 (SVM)	0.863 (RC)	0.877 (SVM)
GaborLBP	0.783 (SVM)	0.849 (Bagging)	0.869 (Bagging)
GLCM	0.909 (SVM)	0.92 (SVM)	0.923 (SVM)
HOG	0.846 (KNN)	0.923 (SVM)	0.909 (SVM)
LBP	0.831 (SVM)	0.886 (Bagging)	0.889 (Bagging)
LQP	0.897 (SVM)	0.929 (SVM)	0.931 (SVM)
LTP	0.9 (SVM)	0.951 (SVM)	**0.954 (SVM)**
MTS	0.903 (SVM)	**0.96 (SVM)**	—
STU1	0.921 (SVM)	0.94 (SVM)	0.949 (SVM)
STU2	0.917 (SVM)	0.929 (SVM)	0.937 (SVM)
WLD	**0.94 (RF)**	—	—

Regarding the classifiers, in almost all the cases the best classifier is the SVM, which is more evident as the number of descriptors concatenated raises. This is because SVM works well with high dimensionality. Our recommendation is to use SVM trained with SMO.

4.2. Backward Subset Selection

The Backward Subset Selection has a similar behavior. Table 4 shows the results of the Backward Subset Selection of descriptors applied to the whole image (BSS1 × 1). This is the case with more iterations. It increases the accuracy from 0.917 to 0.937. The resulting descriptor set is composed by BGC2, BGC3, GLCM, LQP, LTP, MTS, STU1, STU2, and WLD.

Table 4. Backward Subset Selection of descriptors applied to 1 × 1 gridded image (also known as BSS1 × 1). Level 1 uses all the descriptors in set D concatenated. Level 2 uses the concatenation of the descriptors in D without each of the descriptors. The following levels use the concatenation of the descriptors in D without the descriptor that makes score higher of the previous level.

Descriptor	Level 1 D	Level 2 D\	Level 3 D\ GaborLBP +	Level 4 D\ GaborLBP + HOG +	Level 5 D\ GaborLBP + HOG + BGC1 +	Level 6 D\ GaborLBP + HOG + BGC1 + LBP +
BGC1		0.92 (SVM)	0.923 (SVM)	**0.929 (SVM)**	—	—
BGC2		0.914 (SVM)	0.92 (SVM)	0.92 (RF)	0.937 (SVM)	0.931 (SVM)
BGC3		0.917 (SVM)	0.92 (SVM)	0.929 (SVM)	0.934 (SVM)	**0.937 (SVM)**
LBP		0.914 (SVM)	0.92 (SVM)	0.926 (RF)	**0.937 (SVM)**	—
GaborLBP		**0.92 (SVM)**	—	—	—	—
GLCM		0.914 (SVM)	0.914 (SVM)	0.92 (SVM)	0.929 (SVM)	0.923 (SVM)
HOG	0.917 (SVM)	0.903 (SVM)	**0.923 (SVM)**	—	—	—
LQP		0.917 (RF)	0.92 (SVM)	0.923 (SVM)	0.923 (SVM)	0.934 (SVM)
LTP		0.909 (SVM)	0.92 (SVM)	0.92 (RF)	0.926 (SVM)	0.929 (SVM)
MTS		0.914 (SVM)	0.92 (SVM)	0.923 (SVM)	0.929 (SVM)	0.931 (SVM)
STU1		0.917 (SVM)	0.92 (SVM)	0.923 (RF)	0.923 (SVM)	0.926 (SVM)
STU2		0.914 (SVM)	0.92 (SVM)	0.926 (RF)	0.931 (SVM)	0.926 (SVM)
WLD		0.914 (SVM)	0.891 (SVM)	0.82 (RF)	0.834 (RF)	0.934 (RF)

Table 5 shows the results of the Backward Subset Selection of descriptors applied to 4 × 4 gridded images (BSS4 × 4). This time, the improvement is from 0.943 to 0.954. The resulting descriptor set is BGC1 BGC2, BGC3, GaborLBP, GLCM, HOG, LQP, LTP, MTS, STU2, and WLD.

Table 5. Backward Subset Selection of descriptors applied to 4 × 4 gridded image (also known as BSS4 × 4).

Descriptor	Level 1 D	Level 2 D\	Level 3 D \ LBP +	Level 4 D \ LBP + STU1 +
BGC1		0.95 (SVM)	0.951 (SVM)	0.949 (SVM)
BGC2		0.946 (SVM)	0.951 (SVM)	**0.954 (SVM)**
BGC3		0.949 (SVM)	0.95 (SVM)	0.945 (SVM)
LBP		**0.951 (SVM)**	—	—
GaborLBP		0.946 (SVM)	0.951 (RF)	0.949 (RF)
GLCM		0.937 (SVM)	0.937 (SVM)	0.94 (SVM)
HOG	0.943 (SVM)	0.94 (SVM)	0.943 (SVM)	0.946 (SVM)
LQP		0.946 (RF)	0.946 (SVM)	0.946 (SVM)
LTP		0.946 (SVM)	0.951 (SVM)	0.949 (SVM)
MTS		0.946 (SVM)	0.951 (SVM)	**0.954 (SVM)**
STU1		0.946 (SVM)	**0.954 (SVM)**	—
STU2		0.949 (SVM)	0.95 (SVM)	0.949 (SVM)
WLD		0.946 (SVM)	0.946 (SVM)	0.946 (SVM)

Table 6, instead, shows the results of the Backward Subset Selection of descriptors applied to 8 × 8 gridded images (BSS8 × 8). The first iteration achieves and score of 0.92, while in the last iteration the score is 0.946. The improvement is 0.026. The resulting descriptor set is BGC1 BGC2, GLCM, HOG, LQP, LTP, MTS, STU1, STU2, and WLD.

Table 6. Backward Subset Selection of descriptors applied to 8 × 8 gridded image (also known as BSS8 × 8).

Descriptor	Level 1	Level 2	Level 3	Level 4	Level 5
	D	D\	D\ LBP +	D\ LBP + GaborLBP +	D\ LBP + GaborLBP + BGC3 +
BGC1		0.914 (SVM)	0.923 (SVM)	0.943 (SVM)	0.931 (SVM)
BGC2		0.909 (SVM)	0.929 (SVM)	0.934 (SVM)	0.943 (SVM)
BGC3		0.917 (SVM)	0.926 (SVM)	**0.946 (SVM)**	—
LBP		**0.926 (SVM)**	—	—	—
GaborLBP		0.92 (SVM)	0.934 (RF)	—	—
GLCM		0.894 (SVM)	0.9 (SVM)	0.917 (SVM)	0.929 (SVM)
HOG	0.92 (SVM)	0.903 (SVM)	0.914 (SVM)	0.934 (SVM)	**0.946 (SVM)**
LQP		0.906 (RF)	0.917 (SVM)	0.931 (SVM)	0.934 (SVM)
LTP		0.909 (SVM)	0.914 (SVM)	0.929 (SVM)	0.934 (SVM)
MTS		0.909 (SVM)	0.937 (SVM)	0.929 (SVM)	**0.946 (SVM)**
STU1		0.906 (SVM)	0.929 (SVM)	0.934 (SVM)	0.937 (SVM)
STU2		0.914 (SVM)	0.929 (SVM)	0.937 (SVM)	0.943 (SVM)
WLD		0.906 (SVM)	0.923 (SVM)	0.931 (SVM)	0.943 (SVM)

4.3. Comparative between Methods

Figure 7 shows a comparative of the highest scores of each iteration of the different selection techniques. The maximum of each technique is obtained in the previous to the last level.

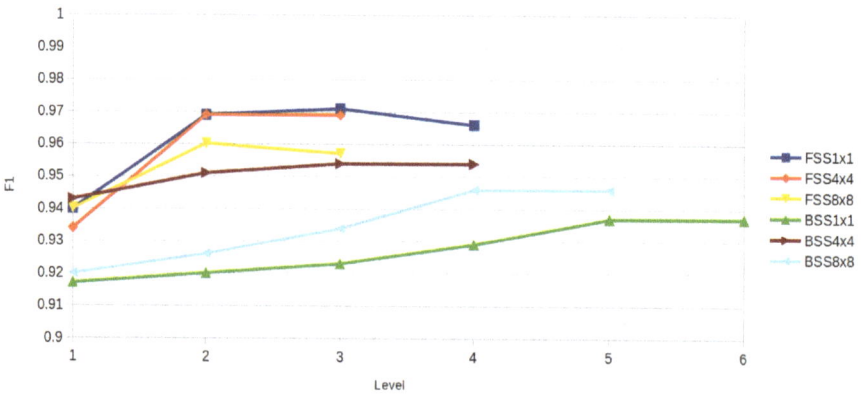

Figure 7. Comparative of highest F1 of each iteration of the Subset Selection techniques. FSS stands for Forward Subset Selection and BSS stands for Backward Subset Selection. The numbers after each selection technique stand for the number of windows the descriptor has been applied to.

On general, Forward Subset Selection achieves better results than Backward Subset Selection. Backward selection computation time is higher than forward so is preferable to use a forward selection since computation time is shorter and the performance is better.

We have compared our method with two known CNN methods: Xception [41] and Siamese [42]. Xception is a Deep learning network inspired by Inception [43], where Inception modules, treated as intermediate step in-between regular convolutions, are replaced by depthwise separable convolutions. Siamese network, instead, is a Convolutional network that inputs two images and classifies if the two images are the same object. One of its advantages is that it gives good results even with small datasets.

Table 7 shows a comparison between the proposed method and the two previous described methods. In the case of the Xception, the results are not as good as Siamese or our proposal as Xception works better for large datasets. Even if Siamese works better than Xception, it does not give better results than our proposal.

Table 7. Comparison between standard DL methods and our proposal.

Method	F1
Xception	0.35
Siamese	0.89
Our proposal ($FSS1 \times 1$)	0.97

In terms of speed, Figure 8 shows a comparative of the test time for each of the methods. The time shown is the average of the different descriptors and classification techniques for testing one image. $FSS1 \times 1$, $FSS4 \times 4$, and $BSS1 \times 1$ have a low computation time, and $BSS8 \times 8$ version performs much slower than the rest of versions due to the high dimensionallity of the data.

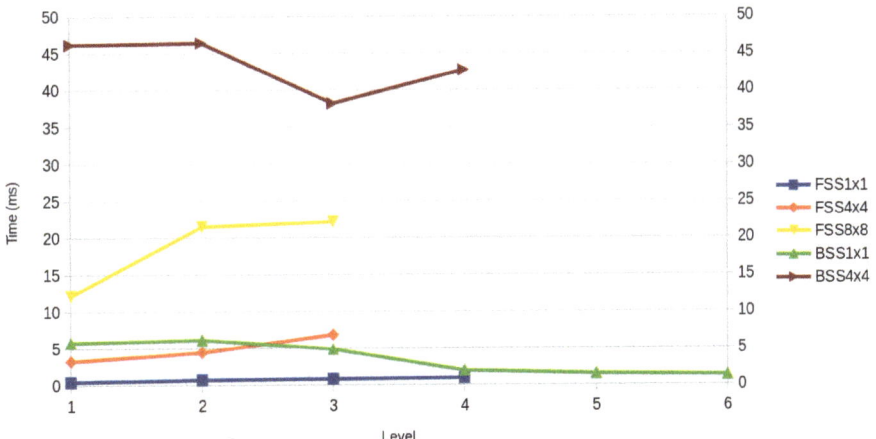

Figure 8. Times to classify an image with the different Subset selection methods on each level. The times on this Figure correspond to the average time that the classifier needs to classify an image using each descriptor on that level. The times of the $BSS8 \times 8$ are not shown since its values are around 200 ms and distorts the plot.

Taking into account the speed and score the best option is to use $FSS1 \times 1$. Although score is similar to the rest of the versions, it outperforms remaining in terms of speed.

5. Conclusions

The main two problems we have to deal with in computer vision in an industrial context is the complexity of the objects, that is, their unusual shape and texture-less objects, and the lack of large datasets to train CNNs that can handle the previous problem. To manage this situation, we proposed in this paper an approach for selecting the best combination of descriptor that, together, provides a better classification. Even if more than one descriptors has to been calculated, this method is still fast enough for real-time applications.

The proposed method is a greedy approach that iteratively adds (Forward Subset Selection) or removes (Backward Subset Selection) descriptors to the solution until performance is not improved. The resulting descriptor set always improves the quality of the classification comparing to the best descriptor by its own. This selection techniques can be extended to different datasets and contexts as proved within this paper and previous ones [34–36].

The used dataset for the experiments is composed by seven typical industrial texture-less objects. The proposed method achieves a state-of-the-art classification quality for that given dataset. Our method achieves a F1 of 0.971, 3% more than the best descriptor alone. Description and classification of a new image can be achieved in real-time applications, given its low processing time (between 10 and 50 ms).

The next steps will include a larger set of descriptors and DL networks in order to mix both classical and Neural Network methods. As this particular application is within a bigger industry 4.0 set-up, the following works will include not only a visual approach, but the application as a whole.

Author Contributions: Conceptualization, I.M., J.A. A.R., and B.S.; methodology, B.S.; software, I.M.; validation, J.A., A.R., and B.S.; formal analysis, I.M.; investigation, I.M.; resources, I.M.; data curation, I.M.; writing—original draft preparation, I.M.; writing—review and editing, I.M., J.A., A.R., and B.S.; visualization, I.M.; supervision, B.S.; project administration, J.A.; funding acquisition, J.A. All authors have read and agreed to the published version of the manuscript.

Funding: This paper has been supported by the project SHERLOCK under the European Union's Horizon 2020 Research & Innovation programme, grant agreement No. 820689.

Conflicts of Interest: The authors declare no conflicts of interest.

Appendix A. Parameters

Table A1. Parameters of the methods.

Algorithm	Parameter	Description of the Parameter	Value
KNN	K	number of neighbors	1
SVM SMO	C	parameter C	1.0
	L	tolerance	0.001
	P	epsilon for round-off error	1.0×10^{-12}
	N	Normalization	true
	V	calibration folds	-1
	K	Kernel	PolyKernel
	C PolyKernel	Cache size of the kernel	250,007
	E PolyKernel	Exponent value of the kernel	2.0
SVM SGD	M	Multiclass type	1-against-all
	F	Loss function	hinge loss
	L	Learning rate	0.001
	R	Regulation constant	0.0001
	E	Number of epochs to perform	500
	C	Epsilon threshold for loss function	0.001
RC	W	The base classifier to be used	RandomTree
	K	Number of choosen attributes in the RandomTree	$int(log_2(predictors)+1)$
	M RandomTree	Minimum total weight in a leaf	1.0
	V RandomTree	Minimum proportion of variance	0.001

Table A1. *Cont.*

Algorithm	Parameter	Description of the Parameter	Value
RF	P	Size of each bag	100
	I	Number of iterations	100
	K	Number of randomly choosen attributes	$int(log_2(predictors) + 1)$
	M RandomTree	Minimum total weight in a leaf	1.0
	V RandomTree	Minimum proportion of variance	0.001
Bagging	P	Size of each bag	100
	I	Number of iterations	10
	W	The base classifier to be used	REPTree (Fast Decision Tree)
	M REPTree	Minimum total weight in a leaf	2
	V REPTree	Minimum proportion of variance	0.001
	N REPTree	Amount of data used for prunning	3
	L REPTree	Maximum depth of the tree	−1 (no restriction)
	I REPTree	Initial class value count	0.0

References

1. Teichmann, M.; Weber, M.; Zoellner, M.; Cipolla, R.; Urtasun, R. MultiNet: Real-time Joint Semantic Reasoning for Autonomous Driving. *arXiv* **2016**, arXiv:cs.CV/1612.07695.
2. Ronneberger, O.; Fischer, P.; Brox, T. U-Net: Convolutional Networks for Biomedical Image Segmentation. *arXiv* **2015**, arXiv:cs.CV/1505.04597.
3. Yan, M.; Zhao, M.; Xu, Z.; Zhang, Q.; Wang, G.; Su, Z. VarGFaceNet: An Efficient Variable Group Convolutional Neural Network for Lightweight Face Recognition. *arXiv* **2019**, arXiv:cs.CV/1910.04985.
4. Liu, Y.; Wang, Y.; Wang, S.; Liang, T.; Zhao, Q.; Tang, Z.; Ling, H. CBNet: A Novel Composite Backbone Network Architecture for Object Detection. *arXiv* **2019**, arXiv:cs.CV/1909.03625.
5. Yuan, Y.; Chen, X.; Wang, J. Object-Contextual Representations for Semantic Segmentation. *arXiv* **2019**, arXiv:cs.CV/1909.11065.
6. Gómez, A.; de la Fuente, D.C.; García, N.; Rosillo, R.; Puche, J. A vision of industry 4.0 from an artificial intelligence point of view. In Proceedings of the 18th International Conference on Artificial Intelligence, Varna, Bulgaria, 25–28 July 2016; p. 407.
7. Chen, C.; Liu, M.Y.; Tuzel, O.; Xiao, J. R-CNN for Small Object Detection. In *Computer Vision—ACCV 2016*; Lai, S.H., Lepetit, V., Nishino, K., Sato, Y., Eds.; Springer International Publishing: Cham, Switzerland, 2017; pp. 214–230.
8. Xie, Q.; Luong, M.T.; Hovy, E.; Le, Q.V. Self-training with Noisy Student improves ImageNet classification. *arXiv* **2020**, arXiv:1911.04252.
9. Cubuk, E.D.; Zoph, B.; Mane, D.; Vasudevan, V.; Le, Q.V. AutoAugment: Learning Augmentation Policies from Data. *arXiv* **2019**, arXiv:1805.09501.
10. Kolesnikov, A.; Beyer, L.; Zhai, X.; Puigcerver, J.; Yung, J.; Gelly, S.; Houlsby, N. Large Scale Learning of General Visual Representations for Transfer. *arXiv* **2019**, arXiv:1912.11370.
11. Wang, J.; Chen, Y.; Yu, H.; Huang, M.; Yang, Q. Easy Transfer Learning By Exploiting Intra-domain Structures. *arXiv* **2019**, arXiv:cs.LG/1904.01376.
12. Nannia, L.; Ghidonia, S.; Brahnamb, S. Handcrafted vs. non-handcrafted features for computer vision classification. *Pattern Recognit.* **2017**, *71*, 158–172.
13. Lowe, D.G. Object recognition from local scale-invariant features. In Proceedings of the Seventh IEEE International Conference on Computer Vision, Kerkyra, Greece, 20–27 September 1999; Volume 2, pp. 1150–1157.
14. Bay, H.; Tuytelaars, T.; Van Gool, L. SURF: Speeded Up Robust Features. In Proceedings of European Conference on Computer Vision, Graz, Austria, 7–13 May 2006; pp. 404–417.
15. Ojala, T.; Pietikäinen, M.; Harwood, D. A comparative study of texture measures with classification based on featured distributions. *Pattern Recognit.* **1996**, *29*, 51–59. [CrossRef]

16. Fernández, A.; Álvarez, M.X.; Bianconi, F. Texture Description Through Histograms of Equivalent Patterns. *J. Math. Imaging Vis.* **2013**, *45*, 76–102. [CrossRef]
17. Merino, I.; Azpiazu, J.; Remazeilles, A.; Sierra, B. 2D Features-based Detector and Descriptor Selection System for Hierarchical Recognition of Industrial Parts. *IJAIA* **2019**, *10*, 1–13. [CrossRef]
18. Haralick, R.M.; Shanmugam, K.; Dinstein, I. Textural Features for Image Classification. *IEEE Trans. Syst. Man Cybern.* **1973**, *SMC-3*, 610–621. [CrossRef]
19. Wang, L.; He, D.C. Texture classification using texture spectrum. *Pattern Recognit.* **1990**, *23*, 905–910.
20. Madrid-Cuevas, F.J.; Medina, R.; Prieto, M.; Fernández, N.L.; Carmona, A. Simplified Texture Unit: A New Descriptor of the Local Texture in Gray-Level Images. In *Pattern Recognition and Image Analysis*; Springer Berlin Heidelberg: Berlin/Heidelberg, Germany, 2003; Volume 2652, pp. 470–477.
21. Xu, B.; Gong, P.; Seto, E.; Spear, R. Comparison of Gray-Level Reduction and Different Texture Spectrum Encoding Methods for Land-Use Classification Using a Panchromatic Ikonos Image. *Photogramm. Eng. Remote Sens.* **2003**, *69*, 529–536.
22. Zhang, W.; Shan, S.; Gao, W.; Chen, X.; Zhang, H. Local Gabor binary pattern histogram sequence (LGBPHS): A novel non-statistical model for face representation and recognition. In Proceedings of the Tenth IEEE International Conference on Computer Vision (ICCV'05) Volume 1, Beijing, China, 17–21 October 2005; Volume 1, pp. 786–791.
23. Tan, X.; Triggs, W. Enhanced local texture feature sets for face recognition under difficult lighting conditions. *IEEE Trans. Image Process.* **2010**, *19*, 1635–1650.
24. Fernández, A.; Álvarez, M.X.; Bianconi, F. Image classification with binary gradient contours. *Opt. Lasers Eng.* **2011**, *49*, 1177–1184.
25. Hussain, S.U.; Napoléon, T.; Jurie, F. Face Recognition using Local Quantized Patterns. In *Proceedings of the British Machine Vision Conference 2012*; British Machine Vision Association: Guildford, UK, 2012; pp. 99.1–99.11.
26. Chen, J.; Shan, S.; He, C.; Zhao, G.; Pietikainen, M.; Chen, X.; Gao, W. WLD: A Robust Local Image Descriptor. *IEEE Trans. Pattern Anal. Mach. Intell.* **2010**, *32*, 1705–1720. [CrossRef]
27. Dalal, N.; Triggs, B. Histograms of oriented gradients for human detection. In Proceedings of the 2005 IEEE computer society conference on computer vision and pattern recognition (CVPR'05), San Diego, CA, USA, 20–25 June 2005; IEEE: Piscataway, NJ, USA, 2005; Volume 1, pp. 886–893.
28. Cover, T.; Hart, P. Nearest neighbor pattern classification. *IEEE Trans. Inf. Theory* **1967**, *13*, 21–27.
29. Rish, I. An empirical study of the naive Bayes classifier. In *IJCAI 2001 Workshop on Empirical Methods in Artificial Intelligence*; IBM: New York, NY, USA, 2001; Volume 3, pp. 41–46.
30. Breiman, L. Random forests. *Mach. Learn.* **2001**, *45*, 5–32. [CrossRef]
31. Cortes, C.; Vapnik, V. Support-vector networks. *Mach. Learn.* **1995**, *20*, 273–297. [CrossRef]
32. Lira, M.M.S.; de Aquino, R.R.B.; Ferreira, A.A.; Carvalho, M.A.; Neto, O.N.; Santos, G.S.M. Combining Multiple Artificial Neural Networks Using Random Committee to Decide upon Electrical Disturbance Classification. In Proceedings of the 2007 International Joint Conference on Neural Networks, Orlando, FL, USA, 12–17 August 2007; pp. 2863–2868.
33. Breiman, L. Bagging predictors. *Mach. Learn.* **1996**, *24*, 123–140. [CrossRef]
34. Shahlaei, M. Descriptor Selection Methods in Quantitative Structure–Activity Relationship Studies: A Review Study. *Chem. Rev.* **2013**, *113*, 8093–8103. [CrossRef] [PubMed]
35. Rasines, I.; Remazeilles, A.; Bengoa, P.M.I. Feature selection for hand pose recognition in human-robot object exchange scenario. In Proceedings of the 2014 IEEE Emerging Technology and Factory Automation (ETFA), Barcelona, Spain, 16–19 September 2014; pp. 1–8.
36. Molina, L.; Belanche, L.; Nebot, A. Feature selection algorithms: A survey and experimental evaluation. In Proceedings of the 2002 IEEE International Conference on Data Mining, Maebashi City, Japan, 9–12 December 2002; pp. 306–313. [CrossRef]
37. Chinchor, N. MUC-4 Evaluation Metrics. In *Proceedings of the 4th Conference on Message Understanding*; Association for Computational Linguistics: Stroudsburg, PA, USA, 1992; pp. 22–29. [CrossRef]
38. Forman, G.; Scholz, M. Apples-to-Apples in Cross-Validation Studies: Pitfalls in Classifier Performance Measurement. *SIGKDD Explor. Newsl.* **2010**, *12*, 49–57. [CrossRef]
39. Platt, J.C. *Sequential Minimal Optimization: A Fast Algorithm for Training Support Vector Machines*; MIT Press: Cambridge, MA, USA, 1998.

40. Robbins, H.; Monro, S. A Stochastic Approximation Method. *Ann. Math. Statist.* **1951**, *22*, 400–407. [CrossRef]
41. Chollet, F. Xception: Deep Learning With Depthwise Separable Convolutions. In Proceedings of the 2009 IEEE Conference on Computer Vision and Pattern Recognition, Miami, FL, USA, 20–25 June 2009; pp. 248–255.
42. Melekhov, I.; Kannala, J.; Rahtu, E. Siamese network features for image matching. In Proceedings of the 2016 23rd International Conference on Pattern Recognition (ICPR), Cancun, Mexico, 4–8 December 2016; IEEE: Piscataway, NJ, USA, 2016; pp. 378–383.
43. Szegedy, C.; Vanhoucke, V.; Ioffe, S.; Shlens, J.; Wojna, Z. Rethinking the Inception Architecture for Computer Vision. In Proceedings of the IEEE Conference on Computer Vision and Pattern Recognition, Las Vegas, NV, USA, 26 June–1 July 2016; pp. 2818–2826.

© 2020 by the authors. Licensee MDPI, Basel, Switzerland. This article is an open access article distributed under the terms and conditions of the Creative Commons Attribution (CC BY) license (http://creativecommons.org/licenses/by/4.0/).

Article

Vision-Based Sorting Systems for Transparent Plastic Granulate

Tadej Peršak [1],*, Branka Viltužnik [2], Jernej Hernavs [1] and Simon Klančnik [1]

[1] Faculty of Mechanical Engineering, University of Maribor, Smetanova ulica 17, 2000 Maribor, Slovenia; jernej.hernavs@um.si (J.H.); simon.klancnik@um.si (S.K.)
[2] Plastika Skaza d.o.o., Selo 22, 3320 Velenje, Slovenia; branka.viltuznik@skaza.com
* Correspondence: tadej.persak@um.si; Tel.: +386-22-207-727

Received: 23 April 2020; Accepted: 18 June 2020; Published: 22 June 2020

Abstract: Granulate material sorting is a mature and well-developed topic, due to its presence in various fields, such as the recycling, mining, and food industries. However, sorting can be improved, and artificial intelligence has been used for this purpose. This paper presents the development of an efficient sorting system for transparent polycarbonate plastic granulate, based on machine vision and air separation technology. The developed belt-type system is composed of a transparent conveyor with an integrated vision camera to detect defects in passing granulates. The vision system incorporates an industrial camera and backlight illumination. Individual particle localization and classification with the k-Nearest Neighbors algorithm were performed to determine the positions and conditions of each particle. Particles with defects are further separated pneumatically as they fall from the conveyor belt. Furthermore, an experiment was conducted whereby the combined performances of our sorting machine and classification method were evaluated. The results show that the developed system exhibits promising separation capabilities, despite numerous challenges accompanying the transparent granulate material.

Keywords: sorting; machine vision; k-NN algorithm; transparent plastic granulate; recycling; air nozzles

---

1. Introduction

Plastic waste is present almost all over the planet and poses serious problems to living organisms and the environment. Such types of waste have decomposition periods in the environment of more than 100 years [1]. Therefore, they should be introduced into the environment as little as possible. Waste plastic decomposes into fragments under five millimeters in size. Such particles are microplastics and are ubiquitous in marine and terrestrial environments, and even in the water we drink [2]. They are also present in other organisms, particularly in marines, which is undesirable. Therefore, plastic waste must be managed correctly to prevent it from reaching the environment [3].

Recycling and reuse are more effective ways to reduce the accumulation of plastic waste, including plastic products that have lost their functionality or have defects from their manufacture. Such products could be of different colors or have some patterns on them, but they can also be transparent. When reusing transparent plastic products, all of the impurities must be removed. Potential impurities have a strong impact on recycled products. They affect their mechanical and visual properties. To this end, sorting machines are used to remove impurities from the mix of used ground plastic [4]. The classification of plastics is essential in the recycling industry, because only in this way can different plastics be separated from one another [5,6]. The sorting machines are of many configurations and operate on different principles for the detection of recycled materials.

There are various techniques for identifying and sorting polymers. Some of these techniques include manual sorting, density separation, electrostatic processes, and various optical systems, including optical inspection using photodiodes or charge-coupled device (CCD) machine vision, near infrared (NIR), ultraviolet (UV), X-ray analysis, and fluorescent light or laser radiation [7].

The most basic non-automated sorting is manual sorting. Manual sorting is prone to error, expensive, tedious, and can be unsafe [7].

Density separation systems are used to separate particles with higher densities than water from buoyant ones. Here, the density of particles must significantly differ [7,8].

Electrostatic separation systems are used to separate a mixture of plastics that can acquire different charges through triboelectrification. It is not suitable for sorting complex mixtures and the particles must be clean and dry [7,9].

Optical systems, which are based on color imaging (Visible light—VIS) sensors can separate different plastics, primarily by color [10]. Problems only arise when the difference between the different colors is very small [11]. In that case, more advanced methods should be used.

Spectrometer and Hyperspectral Imaging are used with a wavelength above a visible range. The former only captures the point and the latter captures the entire line [4]. The most used ranges are VNIR (Visible and near infrared light; 400–1000 nm) and NIR (Near infrared; 800–2500 nm) [10,12]. The purity of the material in reuse is related to the quality of the product. A lot of research has been done to separate the various types of plastics, such as acrylonitrile butadiene styrene (ABS), polystyrene (PS), polypropylene (PP), polyethylene (PE), polyethylene terephthalate (PET), and polyvinyl chloride (PVC). These plastics can be separated from each other completely by spectrography [12,13]. Information on the composition of the material can also be obtained by using spectrometry [14,15]. For better performance, NIR hyperspectral methods are integrated with artificial intelligence methods [16]. These methods are very effective in separating different materials from one another. When looking for differences in the same material with different colors, these methods are less accurate.

X-ray is suitable for the identification of polyvinyl chloride (PVC) from polyethylene terephthalate (PET). X-ray may involve higher system complexity and also health risk [10].

Laser technique offers less-than-a-microsecond fast identification of plastic based on atomic emission spectroscopy. A laser is used to release excited ions and atoms from the material surface, and these can then be identified through spectral analysis to provide polymer type and additives present [7,17,18].

Sorters based on CCD cameras and machine vision occur mainly in sorters of various objects. Support vector machines and artificial neural networks are popular choices for classifiers. Thus, a sorter was developed that separates various objects, such as gears, coins, and connectors [19].

Methods of machine vision, such as deep learning, are emerging in waste sorting. With such methods it is possible to classify and detect various objects, such as glass bottles, paper boxes, paper cups, ceramic plates, and so on [20].

Most of the methods of machine vision in sorting occur in food production. Thus, various defects on crops, such as tomatoes, oranges, lemons, eggs, seeds, and almonds, are inspected with deep learning [21–23], support vector machines [24], neural networks [25], and k-Nearest Neighbors [26].

These methods are also used in the processing of foods, such as fish and chicken. The quality is checked or certain parts of the animal are identified [27,28].

There are also simpler methods for checking seed coloration by counting pixels of a particular color [29].

Researching the sorting machines, it was found that there were different manipulators for the physical separation of particles. These are various pneumatic, electrical, and hydraulic manipulators for sorting larger objects. Ejecting with compressed air and pneumatic nozzles for smaller objects is also widely used [30–32]. There are also advanced algorithms for tracking and simulating single particle excretion. These algorithms are able to track objects on the conveyor belt and separate them from other objects with great precision [33–35].

In this research, the development of a real-time sorting machine for sorting transparent plastic granulate with a lot of different defects was presented. These defects are different black dots, blurs, burns, etc. These products are unusable. This can be done by turning the product into base material for new plastics. In the recycling process, defects caused by product injection molding must be eliminated. For this purpose, a polycarbonate (PC) granulate sorting machine was developed, which removes particles with these defects.

The developed sorting machine made an optical control of the granules with the help of a camera operating in the visible range of light. The particles needed to be inspected, so machine vision was used. Machine vision searched for each individual granule in the image and classified it using the k-Nearest Neighbors (k-NN) method. This method was chosen based on the datatype and application requirements. The images to be processed were of size (40 × 40 pixels), which is very small in comparison to images used in conventional machine vision applications. Granule classification requires real-time processing, so a fast and efficient algorithm is needed. The physical sorting of the granules was performed with air nozzles.

Existing sorters offer the sorting of particles by color and size, and some can even determine the composition of the material the particle is made of. The developed prototype sorter uses machine vision and artificial intelligence methods. With these methods, each individual transparent particle can be classified according to the training database so one can determine whether they are to be ejected. The sorting machine can check their color and any irregularities that may appear in a particle.

2. Materials and Methods

2.1. Materials

2.1.1. Hardware of the Sorting Machine

The development of the sorting machine began in Solidworks 3D modeler. The 3D model was always changing throughout the development, and the final version is presented in Figure 1. The sorting machine hardware consisted of a mechanical part and a control part.

The main part of the mechanical part was a conveyor with a transparent conveyor belt, as shown in Figure 1. The belt was 3 mm thick and made of polyurethane. The width of the belt was 140 mm, the length was 600 mm, and the diameter of the conveyor rollers was 50 mm. The conveyor was driven by a three-phase electro motor with a power of 0.37 kW and a speed of 1370 rpm. Engine speed could be adjusted with the Mitsubishi FR-S520SE-0.2K-EC (Mitsubishi Electric, Tokyo, Japan) inverter. The speed of the belt could be adjusted from 0 to 0.55 m/s.

A feeder oversaw the dosing of the granules to the main conveyor. It consisted of a smaller conveyor on which the granule tank was mounted, as shown in Figure 1. The conveyor was powered by a three-phase electro motor with a power of 0.09 kW and a speed of 1360 rpm. The gearbox ensured the proper speed. Engine speed could be adjusted with the Mitsubishi FR-S520SE-0.2K-EC inverter. The material quantity that was being filled onto the conveyor varied, depending on the speed of the feeder belt and the size of the tank opening.

The material sorting air nozzles consisted of nozzles 1 mm in diameter and 5 mm in spacing, as shown in Figure 1. The air nozzles were connected by pneumatic tubes to a valve block, on which each nozzle had its own pneumatic valve SMC SX11F-GH (SMC Corporation, Tokyo, Japan). The air nozzles were located under the conveyor belt and required a compressed air source and a pressure regulator.

An Industrial Controller managed the entire sorting machine. A camera was mounted above the conveyor to capture the images of material lying on it, as shown in Figure 1. The illumination was positioned below the conveyor in the background configuration, as shown in Figure 1. An encoder with a resolution of 1024 signals per rotation RLS RE22IC0410B30F2C00 (RLS, Pod vrbami, Slovenia) was mounted on the conveyor for position measurement. Both 24 V and 5 V power supplies were required to power the components. A catcher of separated and ejected material, as shown in Figure 1, and adaptive circuits for switching the valves and motors were needed.

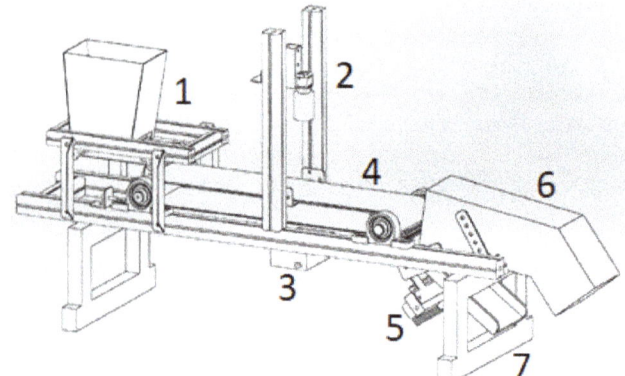

Figure 1. Sorting machine prototype with marked components: 1. Feeder, 2. Camera, 3. Illumination, 4. Conveyor, 5. Air nozzles and valves block, 6. Catcher of ejected material, 7. Catcher of separated material.

Image processing was performed on an Industrial Controller (IC) National Instruments IC-3173 (National Instruments, Austin, Texas, ZDA). The controller had an Intel Core i7-5650U processor; 2.2 GHz, 8 GB RAM. The Industrial Controller also had Field-Programmable Gate Array (FPGA) Xilinx Kintex-7 XC7K160T for the faster processing of lower-level algorithms. A Basler CMOS camera (Basler, Ahrensburg, Germany), model acA2500-14uc was used for image capture. The camera was capable of recording at 14 frames per second at a 2590 × 1942 pixels (5 MP) resolution, and was used with an Opto Engineering 5 Megapixel 12 mm lens.

2.1.2. Samples

The sorting material comprised transparent polycarbonate (PC) in the granules, as shown in Figure 2. The granules were 3 mm wide, 4 mm high, and 2 mm thick. The granules were of different colors. Other defects on the granules were spots that represent burns that can form during the plastic injection process. Through sorting, all of the granules with impurities must be ejected. Only completely transparent material, as shown in Figure 2a, is acceptable for reuse.

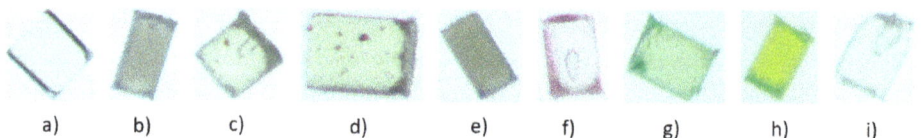

Figure 2. Representatives of each granule class: (**a**) Clean; (**b**) Blur; (**c**) Black dots; (**d**) More black dots; (**e**) Dark; (**f**) Pink; (**g**) Green; (**h**) Yellow (**i**) Material mix.

An image database was created for learning and testing the classifier. All 9 classes are listed in Table 1. For each class, 1000 images of individual granules were made, as shown in Figure 2. This database was used to train and test the classifier.

Table 1. List of used samples.

Number	Class	Accept/Reject
1	Clean	Accept
2	Blur	Reject
3	Black dots	Reject
4	More black dots	Reject
5	Dark	Reject
6	Pink	Reject
7	Green	Reject
8	Yellow	Reject
9	Material mix	Reject

2.2. Methods

2.2.1. Image Acquisition

The camera was triggered by a hardware trigger from the controller. The camera sent the captured image to the controller where the machine vision was executed. A universal serial bus (USB) 3.0 was used for image transfer. The Region of Interest (ROI) size was 57 × 77 mm.

A key part of the machine vision system was lighting. Without consistent illumination quality, images could not be captured. A 48 W light emitting diode (LED) light with color temperature 6000 K was used to illuminate the granules. Granulate illumination is specific, because of reflection on its surface. Therefore, illumination was placed under the conveyor belt in a backlight configuration. Two diffusers were used for more even light. One was mounted directly on the light, the other below the conveyor belt, as shown in Figure 1.

2.2.2. Image Processing

Before processing, the image was captured and stored in a buffer. The images were processed with delay to save time. This means that, while one image was captured, the other was processed. In classic processing, the image would be captured and processed immediately. Thus, the time it took to capture a camera image and transfer the image to a computer was lost. This time would be difficult to save in real-time applications of image processing.

The overall image processing was separated into two main parts. These were the localization and classification of the granules.

At localization, the originally captured images were reduced to 37% of their original size, as shown in Figure 3a, to speed up the image processing time. Then, the RGB green color extraction was performed and the correct brightness was set, as shown in Figure 3b. We then applied a Modified Sauvola threshold [36,37], which makes a binary image with only granules, and some small regions which do not represent granules, as shown in Figure 3c. The processed image was sent to a filter to remove small regions of the image that did not represent granules, as shown in Figure 3d. Finally, the Particle Analysis method from LabVIEW National Instruments software was carried out, which gave us the x and y coordinates of the granules that were in the image. The x and y coordinates of the granules represented the center of mass of the object in the image. The Particle Analysis method takes some time to implement but it is critical to the operation of the sorter.

At the conveyor's full speed, image processing took up to 100 ms. The actual time needed for processing depended on the number of particles in the image. Image capture, granule localization and classification were performed during this time. When a new image capture was triggered, the image capture was started, and the previous image was processed. Thus, the image capture and the processing were performed in parallel. The image buffer and image minimization saved 92 ms in image processing time. Without the implementation of the described procedure, the total image processing time would have been almost 200 ms.

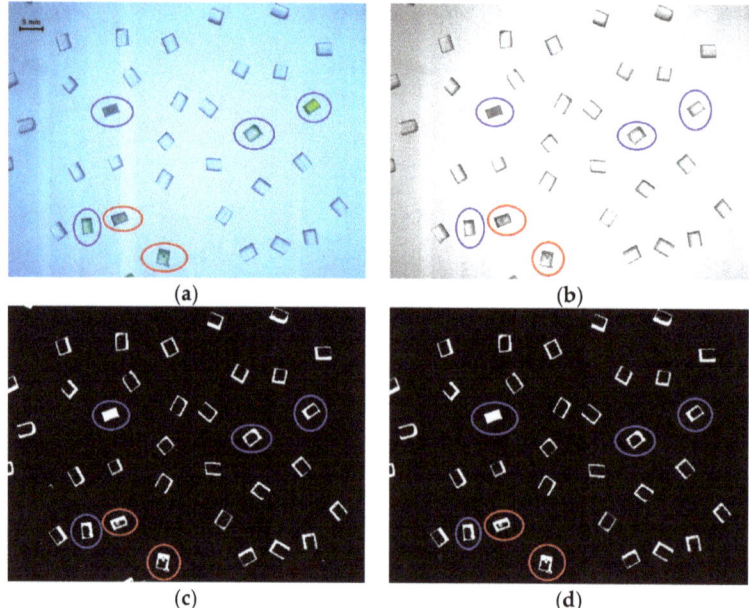

Figure 3. Image processing operations: (**a**) Original image; (**b**) RGB green color extraction and brightness settings; (**c**) Modified Sauvola threshold; (**d**) Particle filter. Blue color indicates marked pellets with color defects and red color indicates pellets with structural defects.

Only granules which were completely transparent and error-free were accepted. Therefore, a color classifier was used. All the granules found in the image were examined by the classifier. The granule location data from Particle Analysis were taken, and the granule image was cut at the granule location and sent to the classifier. The cropped image size was 40 × 40 pixels. Each cropped image was converted to the HSL color space. Then, the hue, saturation, and luminance histograms of the color sample were calculated. The hue and saturation histograms each contained 256 values, and the luminance histogram was reduced to 8 values. Combined, the 520 hue, saturation, and luminance values produced a high-resolution color feature vector. Because very fast real-time processing is required, the dimensions of the feature vector were reduced using a dynamic mask. A reduced color feature vector contains 128 hue and saturation values and 8 luminance values—for a total of 136 values—and represents the input to the classifier. By suppressing the luminance histogram into eight values (12.5% reduction), the algorithm accentuated the color information for the sample.

Color feature vector of granule image was sent to color classifier, which was based on the *k*-NN algorithm. This is a statistical method for classification based on the nearest neighbor [38]. This method was characterized by Lazy Learner, because it does not learn in advance, but only when it receives a classification requirement. When a method receives a classification request, it compares the data obtained with those of its database. It finds the closest or *k*-Nearest Neighbors (*k*-NN) based on Euclidean distance [39]. The accuracy of the *k*-NN classification changes as the number of neighbors change and varies from case to case.

The granules were ejected based on the classification results. If a granule was detected as an error granule, its coordinates were stored in the circular buffer for ejecting.

2.2.3. Evaluating the Classifier

A confusion matrix was used to demonstrate the effectiveness of the classification. Table 2 presents the confusion matrix for classification into several classes. For example, a classifier sorts a problem into specific classes. The output of the classifier may be one of the following possible cases. For example, if particle belongs to the class C2 and is classified as of class C2, the result is a True Positive. If the classifier predicts a C2 particle to be of some other class than C2, the result is a False Negative, designated by β in Table 2. If the classifier predicts a non C2 particle to be of class C2, the result is a False Positive (α) [40,41]. While making predictions on C2 particles, we did not provide data from any other classes, so each particle that was not analyzed and not classified as C2 was a True Negative (γ).

Table 2. Representation of the confusion matrix.

Considering Class C2:		Actual Class				
		C1	C2	C3	C4	C5
PREDICTED CLASS	C1	γ	β	γ	γ	γ
	C2	α	TP	α	α	α
	C3	γ	β	γ	γ	γ
	C4	γ	β	γ	γ	γ
	C5	γ	β	γ	γ	γ

Positives, P: True Positives: TP = designated in the Table; False Positives: FP = $\sum \alpha$. Negatives, N: True Negatives: TN = $\sum \gamma$; False Negatives: FN = $\sum \beta$.

The performance of our classifier was evaluated by three metrics which determined its positive predictive value (1), hit rate (2), and accuracy (3). The precision of the class tells us how many predictions, which the model considered positive, were actually positive. It was calculated as a ratio between True Positives and all positive predictions for the class in question. Ideally, the precise model would never falsely classify other particles to be of the class in question (i.e., no False Positive predictions). However, even an ideally precise model can still make an error of not recognizing that a particle belongs to the certain class. That is the reason for a metric called recall. This tells us how many of class C particles were correctly classified. This was calculated as the ratio between the True Positives and all samples of the class in question. The model with the ideal recall would never make a mistake of falsely predicting a class C particle to be of some other class (i.e., no False Negative predictions). Lastly, we have the metric accuracy, which provided overall prediction quality for each class. It was calculated as a ratio between all true predictions and a total population.

Precision:
$$precision = \frac{TP}{TP + FP} \quad (1)$$

Recall:
$$recall = \frac{TP}{TP + FN} \quad (2)$$

Accuracy for one class:
$$accuracy = \frac{TP + TN}{P + N} \quad (3)$$

Accuracy for the whole confusion matrix:
$$ACC = \frac{\sum True\ Positive + \sum True\ Negative}{\sum Total\ population} \quad (4)$$

2.2.4. Sorting Procedure

The sorting algorithm worked based on a circular buffer. The location information of the granules was obtained from the particle analysis. The coordinate x was used (by the width of the image) to determine which valve should activate and eject the granule. The y coordinate (image length) was used to determine the moment of opening of the pneumatic valve to eject the granule, as shown in Figure 4.

The distance between the granule in the image to be ejected and the nozzle was measured by an encoder. The encoder was mounted on a driven conveyor roller. The belt moves were determined by the encoder pulses. The y coordinate of the granulate, which represented the distance from the image edge, is Y_r. Y_r was converted from the pixel value to the number of encoder pulses and added to a constant Y_e, representing the number of pulses from the image to the air nozzles, as shown in Figure 4. The value of pulses for ejecting Y_s was written to the circular buffer running on the FPGA. The FPGA monitored the encoder value, and opened the air valve for the nozzle, according to the values in the circular buffer.

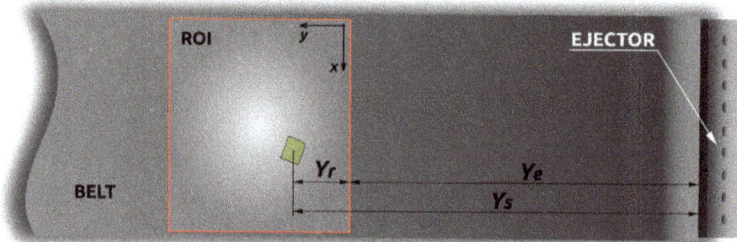

Figure 4. The process of determining which pneumatic valve to open and when to eject the granule.

3. Results and Discussion

The chapter, Results and Discussion, is divided into two parts. The first part presents the in silico results of the classification of the granules into nine classes and into two classes. In the second part, the in vivo results of physical sorting on the prototype sorting machine are presented.

3.1. Classification Results

3.1.1. Classification into Nine Classes

Classification was done with the composite image database. The granulate images were uploaded into a k-Nearest Neighbors algorithm (k-NN) color classifier. The classes were designated by numbers 1–9. The class names are listed in Table 1. A set of 850 learning and 150 testing images were used for each class. Only one nearest neighbor ($k = 1$) was used. With the current image database, the k-NN algorithm with one neighbor got the best results. The same applied for the classification into two classes.

Table 3 shows the result of the classifications. They are represented in the confusion matrix. The columns for each class indicate how many granules were allocated to each class.

Table 3. Confusion matrix of sorting results into nine classes.

	Class	1	2	3	4	5	6	7	8	9
PREDICTED	1	150	0	0	0	0	0	0	0	0
	2	0	145	9	0	3	0	0	0	7
	3	0	2	94	17	7	0	0	0	12
	4	0	0	9	126	10	0	0	0	0
	5	0	1	3	5	128	0	0	0	0
	6	0	0	0	0	0	150	0	0	1
	7	0	0	0	0	0	0	150	0	0
	8	0	0	0	0	0	0	0	150	0
	9	0	2	35	2	2	0	0	0	129
	Σ	150	150	150	150	150	150	150	150	150
	Precision	99.34	88.41	71.21	86.9	93.43	99.34	100	100	75.88
	Recall	100	96.67	62.67	84	85.33	100	100	100	86

Calculation of the classification accuracy into nine classes, according to Equation (3). The calculated average accuracy (ACC) was 90.5%.

The results from Table 3 show that all the granules in class 1 (clean) were classified correctly. In testing other classes, only in one case was the granule recognized as clean, but was not clean. When sorting, it is important to classify clear granules and defective granules precisely, because the sorting machine also separates the material into clean granules and defective granules.

3.1.2. Classification into Two Classes

The sorting algorithm sorts the material into clean granules and defective granules, so, classification was made on clean and defective granules. A set of 850 learning and 150 testing images were used for each class. Only one nearest neighbor ($k = 1$) was used, as with the nine-class classification. Images of clean granules without defects were used in the clean (OK) class. A mix of other images with defectives granules were used in the defective (NOK) class. Table 4 shows the results of the classifications. All the granules were classified correctly.

Table 4. Confusion matrix of results for two classes.

	Class	TRUE OK	NOK
PREDICTED	OK	150	0
	NOK	0	150
	Σ	150	150
	Precision	100	100
	Recall	100	100

OK: clean; NOK: defective.

Calculation of the classification accuracy into two classes was performed using Equation (3). The calculated accuracy was 100%.

Table 4 shows the results of k-NN classifications on clean (OK) and defective (NOK) granules. The classifier was capable of separating granules with 100% accuracy. Because the classification worked with 100% accuracy, any errors that occurred in sorting were the result of other influences. These were the physical effects of the adhesion of the granules to the conveyor and cohesion forces between the granules.

3.2. Sorting Results

The classifier was tested on a prototype sorting machine. A test mixture of granules was prepared, into which clear granules and defective granules were placed. Defective granules represented 10% of the total mixture.

Testing was performed with five different settings for the feeder and the conveyor. The settings are given in Table 5. The parameters are explained in Table 6. Table 7 shows the sorting results.

Table 5. The settings of the feeder and the conveyor when testing the sorting efficiency.

Test Name	S_c—Conveyor Belt Speed (m/s)	S_f—Feeder Belt Speed (m/s)
Test 1	0.545	0.006
Test 2	0.545	0.024
Test 3	0.419	0.013
Test 4	0.308	0.006
Test 5	0.369	0.006

Table 6. Parameters' interpretations in sorting machine testing.

Data	Unit	Equation	Description	Acquisition
t	s		Sorting time	Measurement
m_D_1	g		Weighing of good granules in Box 1	Measurement
S_1	piece		Counting defect granules in Box 1	Measurement
λ		S_c/S_f	Bulk density on conveyor	Calculation
S_0	piece		Bad granules in Box 0	Calculation
D_1	piece		Good granules in Box 1	Calculation
D_0	piece		Good granules in Box 0	Calculation
$m.$	kg/h		Sorting capacity (granular mass flow rate)	Calculation
ni_0		$S_0/(S_0 + D_0)$	The quality of the ejected granules (percentage of defect granules in Box 0)	Calculation
ni_1		$D_1/(D_1 + S_1)$	Quality of separated granules (percentage of good granules in Box 1)	Calculation

Three repetitions of the test with the same sorting machine settings were made with the granule test mixture. Table 7 provides averages of the results of these three tests. The variable parameters were the speed of the conveyor and the speed of the feeder belt. The influence of how densely the granules were arranged on the conveyor was changed by adjusting these two parameters. The size of the opening on the feeder was constant.

Adjusting the conveyor speed also affected how many frames per second should be captured, and in what arc the granules would fall past the air nozzles unit below the conveyor. The faster the conveyor, the faster the image processing should be. At the full speed of the conveyor, image processing took up to 100 ms. This was 10 frames per second, so the camera was fast enough for image capturing.

A schematic presentation of the testing system is shown in Figure 5. The clean granules are represented by the representatives of the clean granules class. Defective granules are a mixture of all other classes. Figure 5 shows two boxes for sorted material. After sorting, Box 1 would ideally contain only clean material. Box 0 contains defective ejected granules.

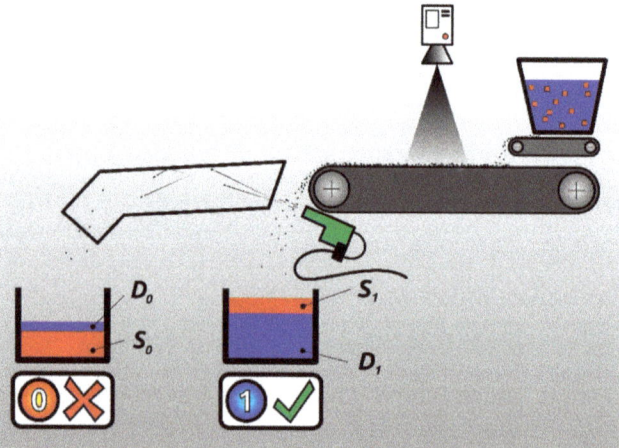

Figure 5. Schematic presentation of the system prototype. The feeder, conveyor and air nozzles are shown. Below are the boxes to catch the material. A "separate" box for non-defective material (Box 1) and an "ejected" box for defective material (Box 0).

Table 7. Sorting results on sorting machine.

Data	Test 1	Test 2	Test 3	Test 4	Test 5
t	236.7	43	65.7	234	222.7
m_D_1	57.5	51.6	55.2	57.6	57.2
S_1	6.3	11.7	45.7	49.7	42
λ	90.8	22.7	32.2	51.3	61.5
S_0	408.7	403.3	369.3	365.3	373
D_1	3324	2983	3191	3330	3307
D_0	203	544	336	197	220
$m.$	1.011	5.567	3.644	1.023	1.075
ni_0	66.81	42.57	52.36	64.97	62.9
ni_1	99.81	99.61	98.59	98.53	98.75

After examining Table 7, the following was determined:

- Sorting speed was measured and depends exclusively on the speed of the feeder belt. The faster the feeder belt rotates, the faster the test is completed.
- The parameter λ tells us how densely granules are arranged on the conveyor. Table 7 shows that they were most densely bulked in test 2, And most sparsely in test 1, which also had the highest sorting quality.
- The sorting capacity (kg/h) depends on the speed of the feeder belt.
- The highest quality of the separated granules (ni_1) and the quality of the ejected granules (ni_0) was achieved with the parameters of test 1—this means at maximum conveyor speed and minimum feeder belt speed. The quality of the separate granules (ni_1) is the most important parameter, because it tells us how clean the material is after sorting.

4. Conclusions

A prototype sorting machine for the rejection of defective plastic granulates has been developed. Research started with capturing images of samples and preparing a training–testing database. There were nine classes in the database. Each class had 850 images to teach and 150 images to test the k-NN classifier. The classification performance in nine grades was 90.52%.

The classification of only two classes was initially carried out. These were defective granules (NOK) and clean granules (OK). Only clear transparent granules were in the OK class. In the NOK class was a mix of defective granules. The classification accuracy, using a k-NN classifier of backlit optical images for the two classes, was 100%. This means that the sorting machine was capable of at least separating the granules theoretically with 100% accuracy.

Particle localization was performed using the Modified Sauvola threshold algorithm. The location of the granulate is important for the operation of the sorting machine, as it is used to send individual granules to the classifier and possibly to eject the granules with air nozzles.

A classifier on a sorting machine was used in the second part. The testing of sorting accuracy was performed on the test samples. The highest purity of the accepted material (defect free) class contained 99.81% pure material (contamination by defective materials was 0.19%).

Classification OK/NOK worked with 100% accuracy, so the conclusion is that all sorting errors are possible due to other influences. These influences can be inaccurate air nozzles separation, error on the determination of granulate location, granule migration on conveyor during moving between camera, and air nozzles and possible software bugs.

The illumination could be more even using better lighting. Better lighting could only improve already good results. The lighting must be very intensive so that the exposure time of the camera can be very short. The speed of the conveyor affects the image quality in the case of too dim lighting and if the camera is rolling the shutter.

Further work could be performed to improve separate ejecting. As the results show, the classification works well, and all errors resulted from the physical manipulation of the granules. Ejecting logic software and hardware could be improved as it could eject the individual granules more accurately. To this end, the possible effects affecting sorting errors should be improved. The main influences are determining the location of the granule and transporting the granules from the feeder to the air nozzles.

Later, the classifier could be adapted for other materials in similar forms. The regrind polycarbonate, which has a very undefined shape, could be also sorted. With this material, the color of the material depends on the particle thickness, which further complicates the classification. The quality of the captured image can also be improved, by improving the illumination and using a telecentric lens. In this way, the images will be of better quality and can determine the location of the granule more precisely. The classification of granules, which is already good, could also be improved.

Other methods of artificial intelligence could be used to classify the granules. These are, for example, neural networks and deep learning. Since a large database for training and testing was made, these methods could be of use, but only major changes to the sorting machine software should be made.

The use of a sorting machine in an industrial environment would be possible. The capacity needs to be increased largely, while maintaining the sorting efficiency. The increase in capacity should follow the example of larger industrial sorting machines, which adjust the sorting capacity with the help of several cameras installed in parallel on the conveyor. So, a wider conveyor should be used and more cameras in parallel should be installed. Depending on the width of the conveyor, the air nozzles should also be adjusted.

Another way to increase the capacity of the sorting machine is the faster movement of the conveyor. However, then the camera must also capture images with more images per second, which also need to be processed. Thus, the speed of image processing must also be increased.

Author Contributions: Conceptualization, T.P. and S.K.; software, T.P. and J.H.; writing—original draft preparation, T.P.; writing—review and editing, T.P., J.H., S.K., and B.V.; project administration, S.K and B.V.; funding acquisition, S.K. and B.V. All authors have read and agreed to the published version of the manuscript.

Funding: This research was funded by the European Regional Development Fund.

Acknowledgments: The authors acknowledge the financial support from the Slovenian Research Agency (Research Core Funding No. P2-0157).

Conflicts of Interest: The authors had no conflicts of interest in this article.

References

1. Chen, X.; Yan, N. A brief overview of renewable plastics. *Mater. Today Sustain.* **2020**, *7–8*, 100031. [CrossRef]
2. Wong, J.K.H.; Lee, K.K.; Tang, K.H.D.; Yap, P.-S. Microplastics in the freshwater and terrestrial environments: Prevalence, fates, impacts and sustainable solutions. *Sci. Total Environ.* **2020**, *719*, 137512. [CrossRef] [PubMed]
3. Li, C.; Busquets, R.; Campos, L.C. Assessment of microplastics in freshwater systems: A review. *Sci. Total Environ.* **2020**, *707*, 135578. [CrossRef]
4. Wu, X.; Li, J.; Yao, L.; Xu, Z. Auto-sorting commonly recovered plastics from waste household appliances and electronics using near-infrared spectroscopy. *J. Clean. Prod.* **2020**, *246*, 118732. [CrossRef]
5. Straka, M.; Khouri, S.; Rosova, A.; Caganova, D.; Culkova, K. Utilization of computer simulation for waste separation design as a logistics system. *Int. J. Simul. Model.* **2018**, *17*, 583–596. [CrossRef]
6. Tange, L.; Van Houwelingen, J.A.; Peeters, J.R.; Vanegas, P. Recycling of flame retardant plastics from weee, technical and environmental challenges. *Adv. Prod. Eng. Manag.* **2013**, *8*, 67–77. [CrossRef]
7. Niaounakis, M. *Biopolymers: Reuse, Recycling, and Disposal*; William Andrew: Amsterdam, The Netherlands, 2013.
8. Wangrakdiskul, U.; Teammoke, P.; Laoharatanahirun, W. Recycled plastic beads sorting machine for polypropylene and acrylonitrile butadiene styrene type with difference of density. *Appl. Mech. Mater.* **2017**, *871*, 230–236. [CrossRef]
9. Zeghloul, T.; Mekhalef Benhafssa, A.; Richard, G.; Medles, K.; Dascalescu, L. Effect of particle size on the tribo-aero-electrostatic separation of plastics. *J. Electrost.* **2017**, *88*, 24–28. [CrossRef]

10. Brunner, S.; Fomin, P.; Kargel, C. Automated sorting of polymer flakes: Fluorescence labeling and development of a measurement system prototype. *Waste Manag.* **2015**, *38*, 49–60. [CrossRef]
11. Spiga, P.; Bourely, A. Application of visible spectroscopy in waste sorting. *Proc. SPIE* **2011**, *8172*, 817212. [CrossRef]
12. Zheng, Y.; Bai, J.; Xu, J.; Li, X.; Zhang, Y. A discrimination model in waste plastics sorting using nir hyperspectral imaging system. *Waste Manag.* **2018**, *72*, 87–98. [CrossRef] [PubMed]
13. Bonifazi, G.; Capobianco, G.; Serranti, S. A hierarchical classification approach for recognition of low-density (ldpe) and high-density polyethylene (hdpe) in mixed plastic waste based on short-wave infrared (swir) hyperspectral imaging. *Spectrochim. Acta Part A* **2018**, *198*, 115–122. [CrossRef] [PubMed]
14. Shameem, K.; Choudhari, K.; Bankapur, A.; Kulkarni, S.; Unnikrishnan, V.; George, S.; Kartha, V.; Santhosh, C. A hybrid libs-raman system combined with chemometrics: An efficient tool for plastic identification and sorting. *Anal. Bioanal. Chem.* **2017**, *409*, 3299–3308. [CrossRef]
15. Juan, H.; Susu, Z.; Bingquan, C.; Xiulin, B.; Qinlin, X.; Chu, Z.; Jinyan, G. Nondestructive determination and visualization of quality attributes in fresh and dry chrysanthemum morifolium using near-infrared hyperspectral imaging. *Appl. Sci.* **2019**, *9*, 1959. [CrossRef]
16. Galdón-Navarro, B.; Prats-Montalbán, J.M.; Cubero, S.; Blasco, J.; Ferrer, A. Comparison of latent variable-based and artificial intelligence methods for impurity detection in pet recycling from nir hyperspectral images. *J. Chemom.* **2018**, *32*, e2980. [CrossRef]
17. Brunnbauer, L.; Larisegger, S.; Lohninger, H.; Nelhiebel, M.; Limbeck, A. Spatially resolved polymer classification using laser induced breakdown spectroscopy (libs) and multivariate statistics. *Talanta* **2020**, *209*, 120572. [CrossRef]
18. Junjuri, R.; Zhang, C.; Barman, I.; Gundawar, M.K. Identification of post-consumer plastics using laser-induced breakdown spectroscopy. *Polym. Test.* **2019**, *76*, 101–108. [CrossRef]
19. Joshi, K.D.; Chauhan, V.; Surgenor, B. A flexible machine vision system for small part inspection based on a hybrid svm/ann approach. *J. Intell. Manuf.* **2020**, *31*, 103–125. [CrossRef]
20. Sousa, J.; Rebelo, A.; Cardoso, J.S. Automation of waste sorting with deep learning. In Proceedings of the 2019 XV Workshop de Visão Computacional (WVC), São Bernardo do Campo, Brazil, 9–11 September 2019; pp. 43–48.
21. da Costa, A.Z.; Figueroa, H.E.H.; Fracarolli, J.A. Computer vision based detection of external defects on tomatoes using deep learning. *Biosyst. Eng.* **2020**, *190*, 131–144. [CrossRef]
22. Nasiri, A.; Omid, M.; Taheri-Garavand, A. An automatic sorting system for unwashed eggs using deep learning. *J. Food Eng.* **2020**, *283*, 110036. [CrossRef]
23. Heo, Y.J.; Kim, S.J.; Kim, D.; Lee, K.; Chung, W.K. Super-high-purity seed sorter using low-latency image-recognition based on deep learning. *IEEE Robot. Autom. Lett.* **2018**, *3*, 3035–3042. [CrossRef]
24. Dhakshina Kumar, S.; Esakkirajan, S.; Bama, S.; Keerthiveena, B. A microcontroller based machine vision approach for tomato grading and sorting using svm classifier. *Microprocess. Microsyst.* **2020**, *76*, 103090. [CrossRef]
25. Narendra Veeranagouda, G.; Amithkumar Vinayak, G. Intelligent computer vision system for vegetables and fruits quality inspection using soft computing techniques. *Agric. Eng. Int.* **2019**, *21*, 171–178.
26. Nasirahmadi, A.; Miraei Ashtiani, S.-H. Bag-of-feature model for sweet and bitter almond classification. *Biosyst. Eng.* **2017**, *156*, 51–60. [CrossRef]
27. Nagaoka, Y.; Miyazaki, T.; Sugaya, Y.; Omachi, S. Automatic mackerel sorting machine using global and local features. *IEEE Access* **2019**, *7*, 63767–63777. [CrossRef]
28. Teimouri, N.; Omid, M.; Mollazade, K.; Mousazadeh, H.; Alimardani, R.; Karstoft, H. On-line separation and sorting of chicken portions using a robust vision-based intelligent modelling approach. *Biosyst. Eng.* **2018**, *167*, 8–20. [CrossRef]
29. Kanjanawanishkul, K.; Chupawa, P.; Nuantoon, T. Design and assessment of an automated sweet pepper seed sorting machine. *Eng. Agric. Environ. Food* **2018**, *11*, 196–201. [CrossRef]
30. Li, H.X.; Li, B.; Choi, J.; Heo, J.; Kim, I. Analysis of a novel nozzle used for pulse jet filtration using cfd simulation method. *Int. J. Simul. Model.* **2016**, *15*, 262–274. [CrossRef]
31. Skews, B.W.; Moss, E.A. Supersonic pulsed jets for material sorting. *Exp. Fluids* **2001**, *31*, 681. [CrossRef]
32. Tourlomousis, F.; Chang, R.C. Dimensional metrology of cell-matrix interactions in 3d microscale fibrous substrates. *Procedia CIRP* **2017**, *65*, 32–37. [CrossRef]

33. Maier, G.; Pfaff, F.; Becker, F.; Pieper, C.; Gruna, R.; Noack, B.; Kruggel-Emden, H.; Längle, T.; Hanebeck, U.D.; Wirtz, S.; et al. Motion-based material characterization in sensor-based sorting. *De Gruyter* **2018**, *85*, 202–210. [CrossRef]
34. Pieper, C.; Pfaff, F.; Maier, G.; Kruggel-Emden, H.; Wirtz, S.; Noack, B.; Gruna, R.; Scherer, V.; Hanebeck, U.D.; Längle, T.; et al. Numerical modelling of an optical belt sorter using a dem–cfd approach coupled with particle tracking and comparison with experiments. *Powder Technol.* **2018**, *340*, 181–193. [CrossRef]
35. Anh, N.T.; Anh, N.H.; Dat, N.T. Development of a framework for ballistic simulation. *Int. J. Simul. Model.* **2018**, *17*, 623–632. [CrossRef]
36. Xiaowei, Z.; Wei, T.; Jianhong, D. A fast adaptive binarization method based on sub block ostu and improved sauvola. In Proceedings of the 2011 7th International Conference on Wireless Communications, Networking and Mobile Computing, Wuhan, China, 23–25 September 2011; pp. 1–5. [CrossRef]
37. Luo, W.; Sun, L. An improved binarization algorithm of wood image defect segmentation based on non-uniform background. *J. For. Res. (1007662X)* **2019**, *30*, 1527. [CrossRef]
38. Han, J.; Kamber, M. *Data Mining: Concepts and Techniques*, 3rd ed.; Morgan Kaufmann: Burlington, MA, USA, 2011.
39. Anton, H. *Elementary Linear Algebra*, 10th ed.; J. Wiley & Sons: Hoboken, NJ, USA, 2010.
40. Tom, F. An introduction to roc analysis. *Pattern Recognit. Lett.* **2006**, *27*, 861–874. [CrossRef]
41. David, M.W.P. Evaluation: From precision, recall and f-measure to roc, informedness, markedness & correlation. *J. Mach. Learn. Technol.* **2011**, *2*, 37–63. [CrossRef]

© 2020 by the authors. Licensee MDPI, Basel, Switzerland. This article is an open access article distributed under the terms and conditions of the Creative Commons Attribution (CC BY) license (http://creativecommons.org/licenses/by/4.0/).

Article

Quality Assessment of 3D Printed Surfaces Using Combined Metrics Based on Mutual Structural Similarity Approach Correlated with Subjective Aesthetic Evaluation

Krzysztof Okarma [1,*], Jarosław Fastowicz [1], Piotr Lech [1] and Vladimir Lukin [2]

1. Department of Signal Processing and Multimedia Engineering, West Pomeranian University of Technology in Szczecin, 70-313 Szczecin, Poland; jfastowicz@zut.edu.pl (J.F.); piotr.lech@zut.edu.pl (P.L.)
2. Department of Information and Communication Technologies, National Aerospace University, 61070 Kharkov, Ukraine; lukin@ai.kharkov.com
* Correspondence: okarma@zut.edu.pl

Received: 30 July 2020; Accepted: 7 September 2020; Published: 9 September 2020

Abstract: Quality assessment of the 3D printed surfaces is one of the crucial issues related to fast prototyping and manufacturing of individual parts and objects using the fused deposition modeling, especially in small series production. As some corrections of minor defects may be conducted during the printing process or just after the manufacturing, an automatic quality assessment of object's surfaces is highly demanded, preferably well correlated with subjective quality perception, considering aesthetic aspects. On the other hand, the presence of some greater and more dense distortions may indicate a reduced mechanical strength. In such cases, the manufacturing process should be interrupted to save time, energy, and the filament. This paper focuses on the possibility of using some general-purpose full-reference image quality assessment methods for the quality assessment of the 3D printed surfaces. As the direct application of an individual (elementary) metric does not provide high correlation with the subjective perception of surface quality, some modifications of similarity-based methods have been proposed utilizing the calculation of the average mutual similarity, making it possible to use full-reference metrics without the perfect quality reference images, as well as the combination of individual metrics, leading to a significant increase of correlation with subjective scores calculated for a specially prepared dataset.

Keywords: additive manufacturing; 3D prints; surface quality assessment; machine vision; image analysis; combined metrics; structural similarity

---

1. Introduction

Additive manufacturing, also referred to as the 3D printing, is one of the key technologies which revolutionizes the small series production in the Industry 4.0 era. The use of the 3D printers makes it possible not only to create some original 3D objects for entertaining purposes but also to launch an individual production of some unique parts of machines and other devices used to replace some damaged older elements. Some other areas of applications of additive manufacturing technology, utilizing plastic filaments usually based on polyactic acid (PLA) or acrylonitrile butadiene styrene (ABS), may be related to biomedical engineering (e.g., individual prosthesis), aerospace and automotive solutions, civil engineering and architecture (e.g., concrete 3D printing), reverse engineering in industry or even the protection of cultural heritage. An integration with 3D scanners enables making custom 3D CAD models and copies of various elements quite easily. Despite a growing popularity of 3D printing for home use, some important limitations should be considered, such as particle emissions [1,2], especially using the ABS filaments.

Nevertheless, the 3D printing process may be affected by various factors influencing the final quality of the manufactured objects, especially considering the low-cost devices designed for home use. Such printers belong to the most popular group based on Fused Deposition Modeling (FDM) where the heated filament is placed from bottom to top layer by layer by the moving extruder forming the 3D printed object. One such source of surface distortions may be the improper melting temperature, dependent on the type of the filament, as well as some changes of the surrounding temperature. Some other issues may be related to poor quality of elements used for printer's construction and low quality of filaments. Some distortions may also be caused by an improper configuration of stepper motors as well as changing filament's delivery speed. Both these types of distortions may be easily forced during the preparation of test samples, which have been used for the development of the database used for verification of the proposed methods also in some earlier papers [3,4].

Increasing popularity and availability of the 3D printers, as well as relatively cheap high-resolution cameras, make it possible to integrate some computer vision algorithms which may be useful for real-time monitoring of the printing process (i.e., without noticeable delays according to the manufacturing speed) and the device state [5,6]. Nevertheless, a majority of known solutions are limited to observation of the device's state and applied mainly for fault diagnosis purposes [7–9]. Some other attempts assume the knowledge of the reference data representing some features or descriptors, such as process signatures [10,11]. Some of the proposed systems utilize optical coherence tomography (OCT) [12], thermographic measurements [13], or terahertz technology [14]. Some other important aspects of detection of some quality issues may be related to cybersecurity of additive manufacturing systems [15,16], especially considering that in some cases some embedded defects in a 3D-printed specimen might remain undetectable, e.g., by ultrasonic inspection [17].

Since the use of sophisticated hardware solutions for monitoring of low-cost 3D printers is troublesome due to a large increase in overall system's costs, the most reasonable solution, particularly for the amateur use, seems to be the analysis of images acquired by affordable cameras. One of such in situ systems for monitoring the manufacturing process based on the comparison of the printed geometry with the computer 3D model has been proposed by Holzmond and Li [18]. Another attempt to anomaly detection and classification for the laser powder bed fusion (LPBF) method based on an unsupervised machine learning algorithm has been proposed by Scime and Beuth [19]. Nevertheless, such automated analysis requires the training of the algorithm, limiting its practical applicability in low-cost devices. A similar problem is also typical for the use of neural networks, adopted e.g., for the monitoring of the 3D inkjet printing process of electronic products [20].

An automated visual inspection of the 3D printing process may also be related to collision detection and the use of visual feedback [21]. On the other hand, some more advanced systems make it possible to optimize the tool paths during manufacturing [22], as well as to use multiple filaments for fabrication [23]. However, the latter solution, known as MultiFab, requires the use of a 3D scanner and closed-feedback loop for small corrections, hence its usefulness for low-cost devices may be limited, even considering the relatively small budget provided by its authors (less than $7000).

An interesting application of machine vision for the detection of defects in 3D prints has been proposed by Straub [24]. This solution, based on five cameras and Raspberry Pi units, has been designed to reduce or eliminate the need for testing of printed objects due to the possibility of automatic correction of minor defects noticed during the printing process, as well as the detection of "dry printing" issues caused by the lack of filament. Nevertheless, the proposed system has turned out to be very sensitive to environmental conditions as well as even minor camera movements. An application of visible sensing for the detection of microdefects in the 3D printed objects used for safety-critical applications has been presented in the paper [25].

Another example of the usefulness of computer vision methods for defect inspection in plastic materials may be the surface control system based on the low-cost visual sensors designed for polyvinyl chloride (PVC) pipes, proposed recently [26].

Regardless of the great progress in the use of machine vision for the monitoring and quality assurance of the 3D printed objects, an automatic visual surface assessment is still a challenging task, particularly correlated with subjective aesthetic perception of the 3D prints. Hence, in this paper, the adaptation of some similarity-based general-purpose full-reference image quality assessment (FR IQA) methods has been proposed together with their combination and optimization towards high correlation with subjective opinions. The experiments have been conducted using a specially prepared database containing 107 images of the 3D printed flat surfaces affected by the various amounts of distortions together with subjective quality scores gathered from 92 human observers.

Although in practical applications the correlation between the proposed objective quality scores and some mechanical properties would be much more desired, the measurement of some of them, e.g., strength of the materials, would require additional destructive experiments. Therefore, in this paper, we focus mainly on the aesthetic quality assessment, similarly as in general-purpose image quality assessment, based on the accordance with subjective opinions. Nevertheless, the developed methods may be further extended, particularly using the data fusion with some other non-destructive measurements, to find some dependencies between them and some physical properties of the manufactured objects.

The rest of the paper is organized as follows: Section 2 contains an overview of similarity-based IQA methods, whereas the description of the developed dataset is provided in Section 3. The idea of the proposed modification of the FR IQA metrics for the quality assessment of the 3D printed surfaces and their combination (Section 4) is followed by the experimental results in Section 5. The paper ends with a discussion (Section 6) and conclusions (Section 7).

2. Overview of Similarity-Based Full-Reference Image Quality Assessment Methods

General-purpose image quality assessment methods may be divided into two major groups, namely subjective and objective methods. The main differences between these two generic approaches are related to the assumed application and required assessment time. The usefulness of the subjective methods in many practical applications is limited by the necessary long time spent on gathering the opinions of individual observers. Hence, in practical applications, objective metrics that may be automatically calculated are more desired. Nevertheless, during the development of objective methods, subjective metrics may be used for verification of their correspondence to human perception of various kinds of image distortions, hence they may be considered mainly in view of aesthetic purposes.

The development of subjective methods is a troublesome and time-consuming task as it requires the engagement of many human observers who fill the questionnaires assessing the quality of the presented images containing various kinds and levels of image distortions. After the statistical analysis of their responses, together with outlier rejection and normalization of scores, it is possible to develop a database of images containing a number of images together with appropriate subjective quality evaluations expressed as Mean Opinion Scores (MOS) or Differential MOS values. During the last years, several such databases have been developed for natural images, as well as for stereoscopic images, video sequences, or even multiply distorted images, which are useful for the verification and performance evaluation of the objective image quality metrics. A comprehensive overview of many IQA databases, as well as various types of quality metrics, may be found, i.e., in the recent survey paper [27].

As the practical applicability of subjective methods is strongly limited by the time necessary to acquire the scores, for many tasks, including optimization of image filtering and lossy compression algorithms, the only possibility is the use of objective metrics which may be calculated automatically without the involvement of the human observers. However, such metrics may be further divided into three families: no-reference metrics (NR IQA), also referred to as "blind" metrics, which do not require the knowledge of the original image, reduced-reference metrics using partial reference information and the most popular full-reference metrics (FR IQA) based on the comparison of the distorted images with the "pristine" reference images without any distortions. Such perfect quality reference images are also included in the IQA databases, helpful mainly in the development of some new FR metrics. Despite the potentially wide areas of applications of "blind" metrics, their universality is still much lower in comparison to FR IQA solutions, causing a higher popularity of the latter approaches.

Most of the general-purpose FR IQA methods are based on the assumption that both compared images represent the same scene but one of the images is corrupted by one or more types of distortions, such as, e.g., the presence of noise, blur, lossy compression, transmission errors, etc. Hence, during the comparison of images, any image registering or spatial adjustment operations, including shifting or rotations are not necessary. Assuming the availability of the reference image, many FR metrics have been proposed during recent years, which utilize the similarity calculation between the original and distorted images. A milestone in the development of such metrics has been the idea of Universal Image Quality Index (UIQI) [28], being the first region-based metric without the disadvantages of the conventional pixel-based metrics, such as Mean Squared Error (MSE) or Peak Signal-to-Noise Ratio (PSNR), known as poorly correlated with subjective quality scores. Due to its potential instability, particularly for black or "flat" (with constant luminance) areas of images, the original version of the UIQI has been shortly replaced by the Structural Similarity (SSIM) [29], where the original formula has been extended as

$$SSIM = l(x,y) \cdot c(x,y) \cdot s(x,y) = \frac{2\bar{x}\bar{y} + C_1}{\bar{x}^2 + \bar{y}^2 + C_1} \cdot \frac{2\sigma_x\sigma_y + C_2}{\sigma_x^2 + \sigma_y^2 + C_2} \cdot \frac{\sigma_{xy} + C_3}{\sigma_x\sigma_y + C_3}, \qquad (1)$$

where the default values of the stabilizing constants (added in comparison to UIQI) for 8-bit grayscale images are: $C_1 = (0.01 \times 255)^2$, $C_2 = (0.03 \times 255)^2$ and $C_3 = C_2/2$. Their role is to prevent the potential division by zero without affecting significantly the overall quality metric's value, calculated as the average similarity value for all positions of the sliding window. The local structural similarity values are calculated as the product of three factors representing luminance distortions $l(x,y)$, contrast loss $c(x,y)$, and structural distortions $s(x,y)$ between the fragments x and y of images A and B, according to the current position of the sliding window. These similarities are based on the local mean values ($\bar{x}$ and $\bar{y}$) of the original and distorted images, the respective variances (σ_x^2 and σ_y^2), and the covariance σ_{xy}. The default window is 11×11 pixels Gaussian window, considered as appropriate for typical standard definition images. Since the SSIM metric has been widely accepted in the community, its implementation has appeared in OpenCV and MATLAB®, for example, and many modifications based on similar formulas have been proposed by various researchers.

Considering the postulated independence of results of the image size without the necessity of changing the sliding window's size, the multi-scale version of this method has been developed by the same authors known as MS-SSIM [30], where the image details at different resolutions are assessed applying iteratively a low-pass filter and downsampling by a factor of 2. The contract and structure comparisons are conducted for each scale, whereas the luminance comparison is computed only for the highest scale. The default normalized weighting exponents have been provided after cross-scale calibration experiments.

During the next several years many other SSIM-based metrics have been proposed, which have also been used in the conducted experiments related to this paper. In 2006 the Quality Index based on Local Variance (QILV) has been proposed [31], where a comparison of the local variance distribution is considered as the extension of the use of mean-variance values in the original SSIM. Finally, the following formula has been proposed, assuming that A and B are the input images:

$$QILV = \frac{2\mu_{V_A}\mu_{V_B}}{\mu_{V_A}^2 + \mu_{V_B}^2} \cdot \frac{2\sigma_{V_A}\sigma_{V_B}}{\sigma_{V_A}^2 + \sigma_{V_B}^2} \cdot \frac{\sigma_{V_A V_B}}{\sigma_{V_A}\sigma_{V_B}}, \qquad (2)$$

where $\sigma_{V_A V_B}$ is the covariance between the variances of two images (V_A and V_B respectively), σ_{V_A} and σ_{V_B} denote the global standard deviations of the local variance with μ_{V_A} and μ_{V_B} being the mean values of the local variance.

Another modification proposed by Sampat et al. [32] is based on the use of the wavelet domain. The metric known as Complex Wavelet Structural Similarity (CW-SSIM) utilizes the assumption that the presence of some distortions causes consistent phase changes in the local wavelet coefficients. The metric is also more robust to small rotations and translations in comparison to the spatial domain SSIM. On the other hand, the application of information content weighting for the multi-scale SSIM (referred to as IW-SSIM), proposed in the paper [33], incorporates the Laplacian pyramid transform to calculate the appropriate weights based on the statistical Gaussian model of natural images. A similar information weighting may also be applied for MSE and PSNR metrics, significantly enhancing their correlation with subjective quality scores.

One of the most successful similarity-based metrics, known as Feature Similarity (FSIM), has been proposed in two versions (for grayscale and color images—denoted as FSIMc) [34] and verified as one of the best metrics for most IQA databases. It utilizes a similar formula as the SSIM but is applied for low-level features, namely phase congruency (PC—as a significance measure of a local structure) and gradient magnitude (GM), a complementary feature extracted using the Scharr filter. For simplicity, the equal importance of these two factors ($\alpha = \beta = 1$) has been assumed in the similarity formula between the images A and B:

$$S_L(\mathbf{x}) = \left(\frac{2 \cdot PC_A(\mathbf{x}) \cdot PC_B(\mathbf{x}) + T_1}{PC_A^2(\mathbf{x}) + PC_B^2(\mathbf{x}) + T_1} \right)^{\alpha} \cdot \left(\frac{2 \cdot GM_A(\mathbf{x}) \cdot GM_B(\mathbf{x}) + T_2}{GM_A^2(\mathbf{x}) + GM_B^2(\mathbf{x}) + T_2} \right)^{\beta}, \quad (3)$$

where T_1 and T_2 are the stability constants preventing the division by zero and $\mathbf{x}$ is the sliding window position. The final formula obtained as the averaged S_L for all locations is additionally weighted by the phase congruency, assuming that the Human Visual System (HVS) is highly sensitive to phase congruent structures and may be expressed as

$$FSIM = \frac{\sum_{\mathbf{x} \in A} S_L(\mathbf{x}) \cdot PC_m(\mathbf{x})}{\sum_{\mathbf{x} \in A} PC_m(\mathbf{x})}, \quad (4)$$

where $PC_m(\mathbf{x}) = max(PC_A(\mathbf{x}), PC_B(\mathbf{x}))$ and $\mathbf{x}$ denotes each position of the local window on the image plane A (or B). Its extension for color IQA utilizes the YIQ color space with two chrominance components (I and Q), used as the two other features, weighted using an additional exponential parameter γ used for tuning the importance of the chromatic data.

A similar approach based on the comparison of gradient images, known as Gradient Similarity (GSM) has been proposed in the paper [35], where the gradient similarity term, calculated in the same way as the luminance or contrast factor in the SSIM, is combined with the luminance comparison (calculated as the $l(x,y)$ term the Formula (1)). As the authors have assumed that the typical Prewitt, Sobel, or Scharr kernels are too small the gradient extraction has been conducted using the following 5×5 pixels mask:

$$Mask = \begin{bmatrix} 0 & 0 & 0 & 0 & 0 \\ 1 & 3 & 8 & 3 & 1 \\ 0 & 0 & 0 & 0 & 0 \\ -1 & -3 & -8 & -3 & -1 \\ 0 & 0 & 0 & 0 & 0 \end{bmatrix}, \quad (5)$$

rotated by 90° to obtain four directional masks for gradient calculations.

Further metrics inspired by the idea of SSIM include a fast and effective spectral residual similarity (SR-SIM), based on specific visual saliency model [36] correlated with perceived image quality, as well as the Edge Strength Similarity (ESSIM) [37], assuming the edge detection based on Scharr kernel and further calculation of edge strength maps determined from the directional derivatives calculated for each pixel. Obtained edge strength maps for the reference and distorted images are compared using the SSIM-like formula, similarly as for the other metrics. The similarity analysis in the frequency domain has led to the idea of the DCT subband similarity (DSS) proposed by Balanov et al. [38]. This approach considers various importance of structural distortions in different DCT subbands from a perceptual point of view. Hence, the similarity values, obtained for individual subbands according to the formula similar to the other mentioned metrics based on the local variances (i.e., their product divided by the sum of squares with stability constants), are weighted providing the final DSS result.

Alternatively, good performance may also be obtained using the SSIM-like formula for measurement of contrast deviation as proposed in the paper [39], where the multiscale contrast similarity deviation (MCSD) has been defined as another relatively fast IQA metric with high accordance with MOS values for the most relevant IQA databases. An interesting fact is that although the contrast has been used in many other metrics, including the original SSIM, in this case, all the calculations are based only on contrast values combined with multi-scale image representation.

An interesting attempt to pooling following the measurement of local distortions has been presented in the paper [40], where the analysis of distortion distribution has been used to improve the performance of the Structural Similarity and Gradient Similarity (the modified methods are referred to as ADD-SSIM and ADD-GSIM, respectively). The authors of this approach have utilized the ranking-based weighting, frequency variation based on structural degradation measurement, and additional entropy gain multiplier.

Recently, some other metrics being the extensions of the original SSIM have been proposed in the paper [41], which utilizes additional predictability of image blocks. The use of the fourth multiplicative component (block predictability) in the metric referred to as SSIM4 is based on the minimal value of mean squared error between the current block and the neighboring ones. Another modification is related to the use of color components (metrics denoted respectively as CSSIM and CSSIM4) and assumes simultaneous computations of metrics for several image scales in the YCbCr color space. Such obtained partial metrics are then averaged with weights obtained by maximizing the rank-order correlations for three datasets.

Two other recent methods utilize contrast combined with visual saliency and Riesz transform with visual contrast. The first one, known as CVSSI (Contrast and Visual Saliency Similarity-Induced Index), has been proposed [42] by the authors of the above mentioned MCSD metric, as the result of the application of the weighted standard deviation of two quality maps obtained for the local contrast and the global visual saliency. The latter approach [43] called Riesz transform and Visual contrast sensitivity-based feature similarity (RVSIM) utilized a Log-Gabor filter for initial image decomposition followed by Riesz transform. Then, the similarity of local amplitude, phase, and direction characteristics is determined forming the similarity matrices subjected to weighting using the visual Contrast Sensitivity Function (CSF), forming a single similarity matrix. It is further combined with gradient magnitude similarity, calculated for the reference and distorted images in the same way as in the FSIM, and subjected to final pooling leading to the overall RVSIM index.

Many of these metrics, as well as some others, have been recently combined using the neural networks for the assessment of remote sensing images [44] leading to promising results. Nevertheless, the direct application of such an approach for the quality assessment of the 3D prints would be troublesome due to the time necessary for the calculation of several metrics as well as the necessity of network training. Considering the variety of the proposed similarity-based approaches to image quality assessment, as well as their relatively short computation time (analyzed also in [44]), the above-described metrics have been examined in this paper to verify their usefulness for an automatic quality evaluation of the 3D printed surfaces, as well as the possibility of their combination, additionally converting them into no-reference methods due to potential applications in real-time systems.

3. The Database of the 3D Printed Surfaces

Verification of the usefulness of various objective IQA metrics for the assessment of the 3D printed surfaces requires the development of a special database containing the images of various distorted and high quality manufactured surfaces together with subjective MOS values. To face this challenge some flat samples have been produced from 9 types of thermoplastic ABS filaments with different colors using three available FDM devices, namely RepRap Ormerod 3, Prusa i3, and XYZprinting da Vinci 1.0 Pro 3-in-1. The choice of the ABS filaments has been motivated by their good mechanical properties and lightness, as well as higher abrasion resistance in comparison with PLA polymers. Nevertheless, relatively high melting temperatures—over 220 °C—together with potentially toxic fumes [1,2] have slightly limited the development of the dataset.

Although the database contains the 107 photographs (together with depth maps obtained by a 3D scanner not used in this paper) of the flat surfaces, the metrics investigated in this paper may be successfully applied regardless of the surface flatness or roundness, hence in this sense, the proposed approach may be considered as universal and may be further verified also for some other surfaces. The images have been acquired in the controlled lighting environment (distributed illumination from three lamps to prevent strong reflections) using a Sony DSC-HX100V camera with the exposure time 1/125 s, 5 mm focal length, and an automatic white balance without flash, ensuring a fixed distance to the surface. The size of such obtained images is 1600 × 1600 pixels, being equivalent to the physical size of the samples equal to 35 mm × 35 mm. Their thickness is about 4 mm and the height of layers varies from 0.3 to 0.35 mm depending on the individual printer and the size of the nozzle [4]. The additional depth maps, however not utilized in this paper, have been acquired using the ATOS 3D scanner manufactured by the GOM company (more information may be found in the paper [3]).

Regardless of the influence of some independent factors, such as quality of filament and construction materials, the presence of distortions in some samples have been forced by changing the temperature, configuration parameters of the stepper motors, filament's delivery speed, as well as using the software model changes. Some of the samples contain cracks as well as the results of under- and over-filling. All the photographs have been independently assessed by 92 human observers using the five-element quality scale from 1 (very poor) to 5 (very good). Obtained results have been averaged to provide the MOS values, being additionally verified by comparisons with previously acquired experts' opinions (used for classification purposes in some earlier works). Some sample images together with MOS values are presented in Figure 1, where the MOS values approaching 5 correspond to perfect quality.

As the primary goal of the paper and the developed database is the proposal of the objective method for the quality assessment of the 3D printed surfaces highly correlated with the subjective perception of surface aesthetics, expressed as the MOS values, the mechanical properties of the samples have not been investigated. Nevertheless, the development of an extended database, useful for the analysis of the 3D structure of the manufactured object, supplemented with the results of terahertz or radiographic detection and identification of defects is planned as a part of the future work, considering also some non-planar objects.

Figure 1. Sample images from the developed database with their average subjective quality scores sorted from the highest to lowest: (**a**) high quality salmon pink sample (MOS = 4.7253), (**b**) high quality pink sample (MOS = 4.6923), (**c**) high quality brown sample (MOS = 4.1333), (**d**) moderately high quality red sample (MOS = 2.4130), (**e**) moderately low quality yellow sample (MOS = 1.4130), (**f**) low quality dark green sample (MOS = 1.1868), (**g**) low quality black sample (MOS = 1.0978), (**h**) low quality fluorescent pink sample (MOS = 1.0110), (**i**) low quality blue sample (MOS = 1.0000).

4. Proposed Approach

The idea of the combination of various features is partially utilized in many metrics discussed in Section 2, starting from the Structural Similarity, being in fact a combination of three components, representing the luminance, contrast, and structure. Nevertheless, such combination may also be applied for individual (elementary) metrics instead of features, leading to a significant increase in the correlation with subjective quality evaluations. The most advantageous results may be obtained for the combination of metrics utilizing various kinds of features or different domains, which are somewhat complementary to each other.

The first such attempt for the general-purpose IQA, based on the nonlinear combination of three metrics, similarly as in the construction of the SSIM or FSIM formulas, has been proposed in the paper [45], where the weighted product of MS-SSIM [30], Visual Information Fidelity (VIF) [46] and R-SVD metric [47] has been used. A modified metric referred to as Combined Image Similarity Index (CISI) [48] has been presented assuming the replacement of the R-SVD metric with FSIMc. According to one of the recent papers [49], presenting a comprehensive overview and comparison of many IQA methods, this metric (actually based on two similarity-based formulas) seems to be still competitive with many recently proposed more sophisticated FR IQA approaches, also for multiply distorted image databases, confirming the validity of the assumptions made in this work as well. Hence, the idea of combination of similarity-based metrics is considered as worth investigating also for the 3D printed surfaces.

Some other approaches to multi-method fusion have also been proposed by some other researchers, e.g., based on the regression approach [50], also with the use of machine learning [51], application of neural networks [52], also for remote sensing images [44], or genetic algorithms [53]. Nevertheless, considering the assumed application for the quality assessment of the 3D printed surfaces during the manufacturing process, the use of a limited number of metrics combined using their weighted product is assumed according to the general formula:

$$Q_{combined} = \prod_{n=1}^{N} Metric_n^{weight_n}, \qquad (6)$$

where N is the number of the weighted elementary metrics.

As the original similarity-based metrics, starting from UIQI and SSIM, belong to the group of full-reference IQA methods, their direct application for the considered tasks would require the knowledge of the reference image. In practical applications, it is usually impossible and even the use of the rendered model of the 3D printed object would require a precise image registering and phase adjustment to compare two images in the same way as for the general-purpose IQA. An additional problem might be related fo different colors and luminance of the acquired images and those rendered from the 3D models. Hence, the idea of the calculation of the average mutual similarity is proposed for the images of the 3D printed surfaces assuming the side location of cameras, initially investigated in the papers [54,55] exclusively for classification purposes. Such side-view mounting of a camera makes it possible to acquire images with visible individual layers of the filament, as illustrated in Figure 1.

The idea of the calculation of the mutual similarity requires the division of the image representing the manufactured sample into regions and the application of the IQA formula for each pair of such obtained blocks. In the conducted experiments the division into 4, 9, and 16 square blocks has been assumed, although the best results have been obtained for 16 blocks (grid of 4×4 regions) requiring 120 mutual comparisons. Since the division of the surface into regions is fixed, we may expect the same or very similar orientations of the patters in each region. Hence, due to the use of the mutual similarity based on the comparison of regions, a potential influence of the position and the orientation of the cropping regions would be significantly reduced. Due to the use of the sliding window approach in the SSIM-based quality metrics, the structural changes introduced by such relatively small rotations or translations influence the final results of individual (elementary) metrics not as much as the presence of physical distortions of the 3D printed surface. Additionally, a phase-shifting by a few pixels to adjust the phase of the compared patterns may also be applied as presented in one of our earlier papers [56]. The influence of the color of individual samples has been reduced using the color to grayscale conversion according to ITU Recommendation ITU-R BT.601-7 [57] (using the *rgb2gray* function in MATLAB®). The illustration of the idea of the mutual similarity is presented in Figure 2, where two examples for the division into 4 and 9 regions are presented with the necessary 6 and 36 mutual similarity calculations. The number of similarities may be expressed as

$$k = \frac{M \cdot (M-1)}{2}, \qquad (7)$$

where M is the number of regions.

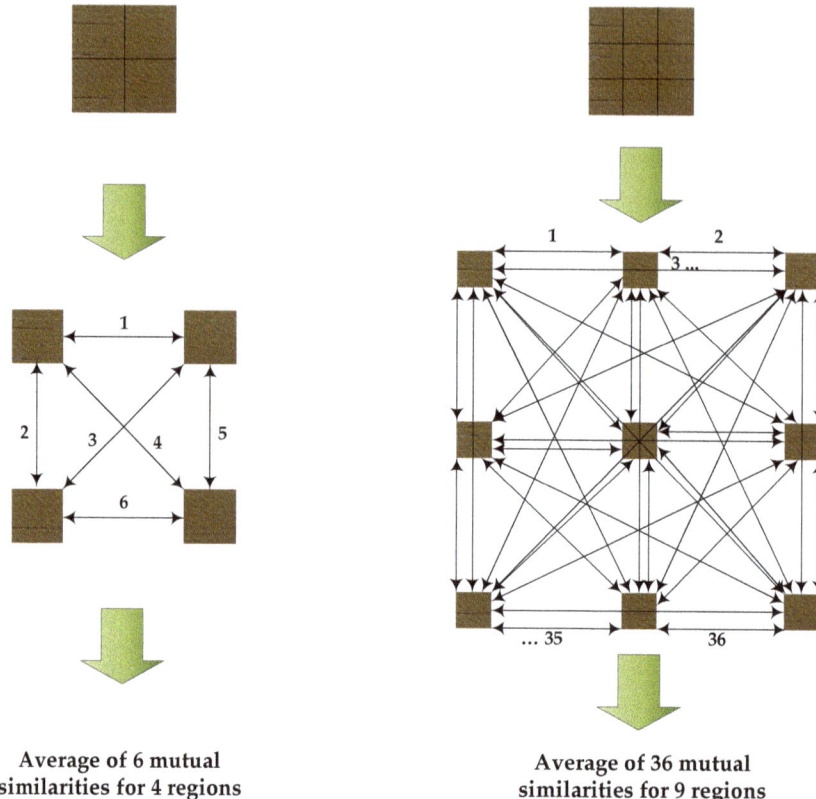

Figure 2. Illustration of the idea of mutual similarity calculation for the division into 4 and 9 regions with necessary respective 6 and 36 computations of the FR metric.

Therefore, the average mutual similarity values obtained for the elementary metrics may be considered as the no-reference equivalents of the FR IQA metrics applied for the quality assessment of the 3D printed surfaces. Since we do not use the "pristine" reference images, we cannot apply FR IQA methods directly. Therefore, we have divided the images into parts (regions) and compare these regions using the FR IQA methods to make a workaround of the lack of the reference images. Hence, formally we can classify our approach as an NR ("blind") IQA approach, although all metrics used internally belong to the group of FR methods.

Such obtained quality scores have been used as the input metrics for the combinations according to Formula (6). The optimized weights have been obtained by the maximization of the Pearson's Linear Correlation Coefficient (PLCC) between the objective and subjective scores (expressed as MOS values in the developed database), representing the prediction accuracy. Since it has been assumed that the nonlinear combination of elementary metrics should compensate the potentially nonlinear perception of distortions, no fitting functions have been used, differently than for the general-purpose IQA being typically applied according to recommendations of the Video Quality Experts Group (VQEG). Additionally, two rank order correlations have been calculated, representing the prediction monotonicity, namely Spearman Rank Order Correlation Coefficient (SROCC) marker as ρ and Kendall Rank Order Correlation Coefficient (KROCC) denoted as τ. They are defined as

$$\rho = 1 - \frac{6 \cdot \sum d_i^2}{n \cdot (n^2 - 1)} \qquad (8)$$

and

$$\tau = \frac{n_c - n_d}{0.5 \cdot n \cdot (n-1)} \qquad (9)$$

respectively, where n is the number of images, n_c and n_d are the numbers of concordant and discordant, and d_i is the difference between the position of the i-th image in two sequences ordered according to subjective and objective scores. Both rank-order correlations consider only the positions of the images in these sorted vectors ignoring the differences between the perceived and measured quality.

5. Experimental Results

The procedure of experiments consists of four main steps: production of test samples with various surface quality, subjective perceptual experiments and data analysis, calculation of individual IQA metrics using the assumed division into regions, and combination of metrics and optimization of weights, as illustrated in Figure 3.

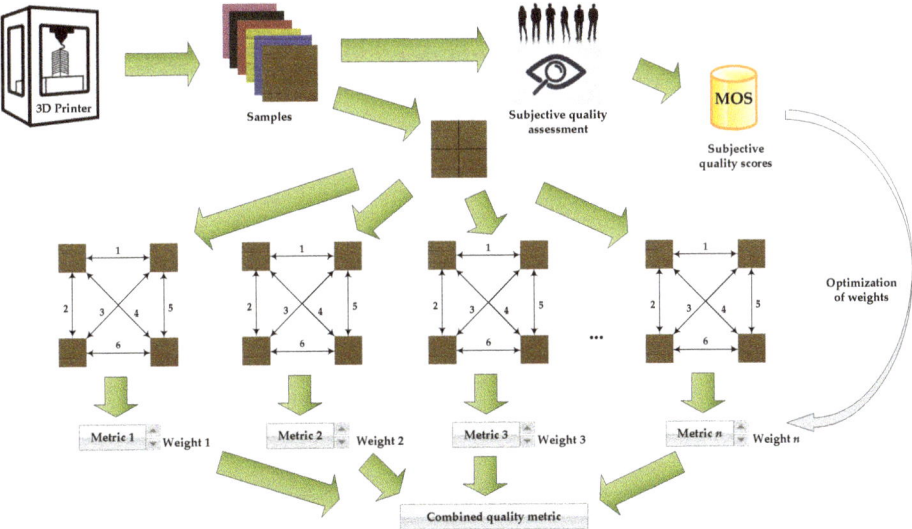

Figure 3. Illustration of the experimental procedure.

The first experiments have been conducted assuming the use of individual elementary metrics for all 107 images converted to grayscale and divided into 4, 9, and 16 regions. For each of them, three correlation coefficients have been calculated to identify potentially the most useful metrics for further combinations. The obtained results are presented in Table 1. As may be noticed, in most cases similar values of PLCC and SROCC have been obtained for individual metrics, hence in further experiments, only Pearson's correlation has been used as the optimization criterion. For many metrics the choice of the number of regions does not change the correlation values significantly, hence further experiments have been conducted assuming the division into 16 regions. The best PLCC results obtained for various combinations of two metrics are presented in Table 2 (to save space only the values higher than 0.67 have been presented). As may be seen, in most combination the FSIM has been used, hence it has been used as the "fixed" metric for the combinations of three metrics (only the two others have been changed during the optimization of weights, although the FSIM exponent has been optimized as well). The best PLCC results achieved for such combinations of three metrics are presented in Table 3.

Table 1. The correlation coefficients obtained for 107 images using the proposed mutual similarity-based approach for the elementary metrics.

Metric	Division into 16 Regions			Division into 9 Regions			Division into 4 Regions		
	PLCC	SROCC	KROCC	PLCC	SROCC	KROCC	PLCC	SROCC	KROCC
FSIM [34]	0.6756	0.6865	0.5195	0.6820	0.6845	0.5185	0.6780	0.6826	0.5114
CW-SSIM [32]	0.5929	0.5823	0.4028	0.6323	0.6098	0.4232	0.5807	0.5633	0.3981
AD-GSIM [40]	0.4081	0.3515	0.2453	0.4020	0.3470	0.2414	0.3873	0.3324	0.2354
DSS [38]	0.4066	0.3523	0.2411	0.3842	0.3220	0.2210	0.3921	0.3176	0.2142
AD-SSIM [40]	0.4017	0.3574	0.2562	0.3834	0.3209	0.2270	0.3492	0.2932	0.2065
SSIM [29]	0.3996	0.4012	0.2746	0.3905	0.4039	0.2661	0.3048	0.3017	0.1938
CSSIM4 [41]	0.3596	0.3296	0.2354	0.3329	0.2818	0.1977	0.3233	0.2851	0.1991
IW-SSIM [33]	0.3549	0.3669	0.2619	0.3230	0.2997	0.2044	0.3169	0.2473	0.1627
SR-SIM [36]	0.3173	0.2441	0.1588	0.3174	0.2497	0.1652	0.3878	0.3160	0.2150
MCSD [39]	0.3106	0.2958	0.2164	0.3008	0.2889	0.2090	0.2952	0.2825	0.2051
QILV [31]	0.3092	0.1330	0.0868	0.3478	0.2662	0.1832	0.4316	0.3555	0.2534
CVSSI [42]	0.2558	0.2083	0.1370	0.2097	0.1492	0.0935	0.1667	0.1068	0.0593
ESSIM [37]	0.1865	0.2340	0.1631	0.1754	0.2354	0.1648	0.3160	0.2868	0.2026
CSSIM [41]	0.1523	0.1078	0.0724	0.1251	0.0755	0.0519	0.1293	0.0862	0.0632
SSIM4 [41]	0.1283	0.0852	0.0565	0.1031	0.0447	0.0304	0.1085	0.0673	0.0462
GSM [35]	0.1103	0.1689	0.1182	0.0991	0.1631	0.1133	0.2102	0.2253	0.1585
RVSIM [43]	0.0267	0.0198	0.0219	0.0546	0.0433	0.0395	0.0114	0.0247	0.0304

Table 2. The PLCC values obtained for 107 images using 25 "best" combinations of two elementary metrics assuming the division into 16 regions.

Metrics	PLCC	Metrics	PLCC	Metrics	PLCC	Metrics	PLCC
FSIM + MCSD	0.8192	FSIM + SR-SIM	0.7862	FSIM + SSIM4	0.7581	FSIM + RVSIM	0.6875
FSIM + DSS	0.8029	FSIM + AS-SSIM	0.7861	FSIM + ESSIM	0.7377	CSSIM + CSSIM4	0.6852
FSIM + CVSSI	0.8008	FSIM + CW-SSIM	0.7766	FSIM + GSM	0.7312	FSIM + SSIM	0.6809
FSIM + IW-SSIM	0.7981	FSIM + CSSIM4	0.7719	SSIM + AD-GSIM	0.7081	FSIM + RVSIM	0.6762
FSIM + AD-GSIM	0.7921	FSIM + CSSIM	0.7593	FSIM + QILV	0.6952	SSIM + CW-SSIM	0.6731

Table 3. The PLCC values obtained for 107 images using 24 "best" combinations of three elementary metrics assuming the division into 16 regions.

Metrics	PLCC	Metrics	PLCC	Metrics	PLCC
FSIM + CW-SSIM + MCSD	0.8472	FSIM + MCSD + GSM	0.8270	FSIM + MCSD + AD-SSIM	0.8221
FSIM + CW-SSIM + DSS	0.8379	FSIM + ESSIM + MCSD	0.8256	FSIM + MCSD + CSSIM	0.8221
FSIM + CW-SSIM + IW-SSIM	0.8356	FSIM + CW-SSIM + CSSIM4	0.8250	FSIM + CVSSI + GSM	0.8219
FSIM + CW-SSIM + CVSSI	0.8348	FSIM + CW-SSIM + SR-SIM	0.8246	FSIM + MCSD + RVSIM	0.8217
FSIM + AD-SSIM + SR-SIM	0.8341	FSIM + CW-SSIM + AD-SSIM	0.8239	FSIM + DSS + AD-SSIM	0.8213
FSIM + CW-SSIM + AD-GSIM	0.8301	FSIM + MCSD + SR-SIM	0.8238	FSIM + ESSIM + AD-GSIM	0.8213
FSIM + DSS + ESSIM	0.8284	FSIM + MCSD + SSIM4	0.8225	FSIM + MCDS + QILV	0.8201
FSIM + DSS + GSM	0.8274	FSIM + MCSD + SSIM	0.8224	FSIM + AD-GSIM + GSM	0.8199

In view of results presented in Table 3, the best results may be obtained using the combination of FSIM and CW-SSIM with a third metric (preferably MCSD), however good results may also be achieved using FSIM and MCSD as two basic metrics used in the combinations. Considering these results, the application of more metrics has led to an even slightly better correlation with MOS values as presented in Table 4, where the obtained rank order correlations are additionally presented as well for comparison with Table 1. As may be seen, the best PLCC values are not always equivalent to the highest SROCC and KROCC, although the differences may be considered as negligible. The additional illustration of the obtained increase of the correlation with subjective scores is presented in the scatter plots in Figure 4.

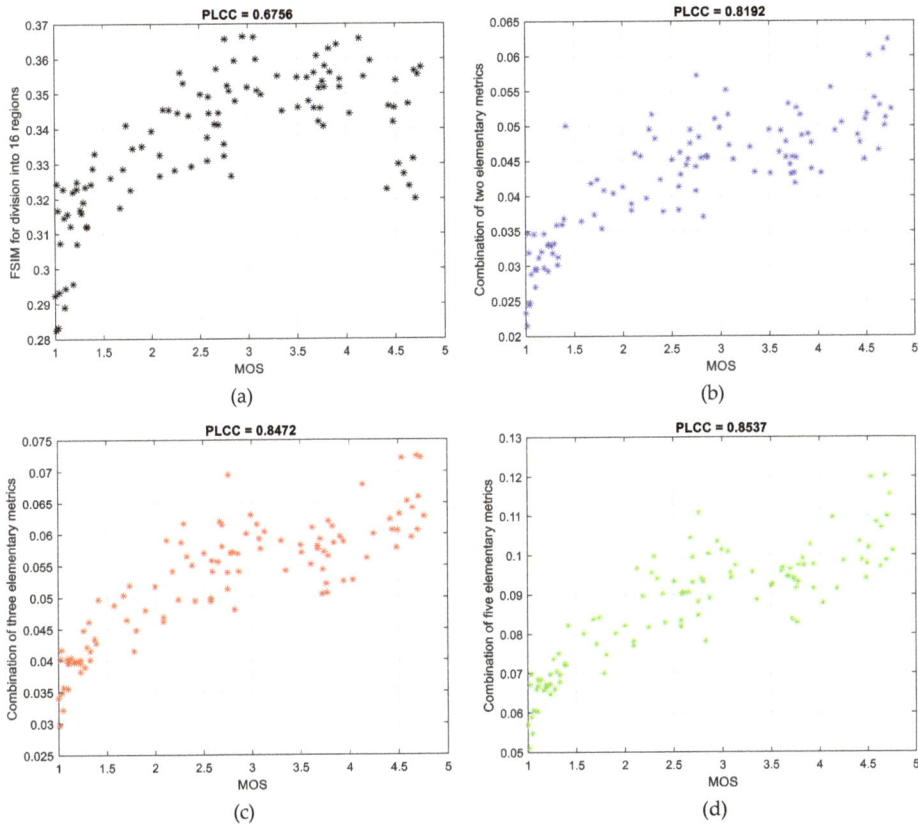

Figure 4. Scatter plots illustrating the linearity of relations between the investigated objective metrics assuming the division into 16 regions and MOS values: (**a**) for FSIM, (**b**) for the "best" combination of two metrics, (**c**) for the "best" combination of three metrics, (**d**) for the "best" combination of five metrics.

Although the verification of correlations for the combinations of two or three metrics calculated assuming the division into 4 or 9 blocks has confirmed the validity of the choice of 16 regions (leading to better performance), the combination of four and five metrics for a smaller number of blocks gives an opportunity to obtain better results, particularly considering the prediction monotonicity represented by both rank order correlations, significantly decreasing the computation time as shown in Table 4.

Table 4. The correlation coefficients obtained for 107 images using the "best" combinations of four and five elementary metrics assuming the division into 4, 9, and 16 regions.

Number of Regions	Metrics	PLCC	SROCC	KROCC
4	FSIM + CW-SSIM + MCSD + ESSIM	0.8405	0.8431	0.6639
	FSIM + CW-SSIM + MCSD + GSM	0.8474	0.8485	0.6710
	FSIM + CW-SSIM + MCSD + ESSIM + GSM	0.8565	0.8586	0.6826
9	FSIM + CW-SSIM + MCSD + ESSIM	0.8463	0.8357	0.6558
	FSIM + CW-SSIM + MCSD + GSM	0.8487	0.8385	0.6565
	FSIM + CW-SSIM + MCSD + ESSIM + GSM	0.8495	0.8397	0.6604
16	FSIM + CW-SSIM + MCSD + ESSIM	0.8502	0.8453	0.6660
	FSIM + CW-SSIM + MCSD + GSM	0.8514	0.8443	0.6653
	FSIM + CW-SSIM + MCSD + ESSIM + GSM	0.8537	0.8480	0.6748

6. Discussion

Analyzing the results, obtained by the unconstrained nonlinear optimization od exponent weights using the MATLAB® *fminsearch* function, which utilizes the simplex search method, additionally verified using gradient-based methods, it may be noticed that the application of an individual elementary metric leads to PLCC and SROCC values not exceeding 0.7 for the best metric (FSIM) regardless of the chosen division into regions. Nevertheless, the application of the combination of even two metrics increases the Pearson's correlation significantly reaching the values over 0.8 for three "best" combinations. However, removing the FSIM metric from the combination decreases this value to only 0.7081, hence Feature Similarity should be considered as the basic metric used in all combinations.

Further increase of the PLCC for the combination of three metrics allows one to specify the other metrics leading to even better performance (CW-SSIM and MCSD, used also in further combinations of four and five metrics) reaching 0.8472 with further increase to 0.8537 for five metrics, also leading to a substantial improvement of the rank order correlations. The increase of the linearity of the relationship between the combined metrics and MOS values may also be observed in Figure 4, particularly for combinations of 3 and 5 elementary metrics. As presented in Table 4, even better results may be achieved limiting the number of blocks for the combinations of four and five individual metrics, obviously with additional advantageous shortening of the overall computation time.

In comparison to previous studies, the obtained results are superior to those presented in the recent conference paper [58], where PLCC = 0.8353 has been obtained for the combination of various approaches, previously applied only for sample classification purposes. However, the methods applied in the aforementioned paper, apart from the use of FSIMc metric, are based on various additional calculations including preprocessing with the Contrast Limited Adaptive Histogram Equalization (CLAHE) algorithm, entropy calculations [3,4], use of the Histogram of Oriented Gradients (HOG) and—more importantly—entropy of the depth maps obtained using the additional 3D scanning.

Considering the practical applicability of the proposed approach, the use of even a monochrome camera is effortless, being enough to acquire the necessary data without the need of the time-consuming 3D scanning of the samples, which makes the real-time applications almost impossible, or at least expensive and troublesome, especially in low-cost solutions.

7. Conclusions and Further Work

Presented results confirm the usefulness of the combined IQA metrics for the quality evaluation of the 3D printed surfaces, leading to a high correlation with subjective opinions of the surface aesthetics. The proposed approach to automatic objective quality evaluation of the 3D printed surfaces is based only on the full-reference image quality metrics that rely on the mutual similarity, hence its application is relatively easy since the values of the elementary metrics necessary for the combination may be calculated in parallel. The utilized individual metrics may be effectively calculated [44,49] and therefore the presented approach may be successfully implemented for the real-time quality monitoring during the manufacturing, understood as not introducing noticeable delays in comparison to the relatively slow printing process. An important remark in view of the computation time is the possibility of an increase in the correlation with MOS values using more metrics but decreasing the number of regions.

Application of the proposed combined metrics makes it possible to increase the correlation with the subjective quality evaluation of the 3D printed surfaces by over 0.17 in comparison to the use of the best single metric, considering the prediction accuracy. A similar increase may be observed for SROCC with the increase of KROCC by more than 0.15, providing significantly higher prediction monotonicity.

As the proposed method is color independent, the use of a monochrome camera is enough, although the application of some other color to grayscale conversion methods as well as various color spaces is planned in further research. Another direction of future work is the extension of the proposed method towards the use of some other image quality assessment methods and integration with previously developed methods used so far for classification purposes. Nevertheless, the development of even better-correlated metrics requires additional experiments, as well as the possible extension of the developed dataset using also some additional samples with non-planar surfaces (with the time-consuming acquisition of subjective quality scores for the new samples). Such an extension of the proposed method for the evaluation of non-planar surfaces is possible similarly as for previously investigated entropy-based approaches [4]. Obviously, reasonable results may be expected only for the division into relatively small regions, additionally avoiding the mutual similarity calculations for distant image fragments due to expected different orientation of both compared to local patterns.

Considering some open challenges of the quality control in 3D printing manufacturing analyzed in the papers [59,60], another direction of further planned experiments is related to the integration of the proposed vision-based methods with electromagnetic and radiographic non-destructive testing (NDT) methods, e.g., using terahertz methods, towards the acquisition of the full 3D structure of the manufactured objects in view of their mechanical properties.

Author Contributions: Conceptualization, J.F. and K.O.; methodology, K.O. and V.L.; software, J.F. and K.O.; validation, P.L. and K.O.; formal analysis, P.L., K.O., and V.L.; investigation, J.F., K.O., and V.L.; resources, J.F. and V.L.; data curation, J.F. and P.L.; writing—original draft preparation, J.F. and K.O.; writing—review and editing, V.L.; visualization, J.F. and P.L.; J.F. worked under the supervision of K.O. All authors have read and agreed to the published version of the manuscript.

Funding: The research is partially co-financed by the Polish National Agency for Academic Exchange (NAWA) and the Ministry of Education and Science of Ukraine (agreement with KhAI no. M/45-2020) under the project no. PPN/BUA/2019/1/00074 entitled "Methods of intelligent image and video processing based on visual quality metrics for emerging applications".

Conflicts of Interest: The authors declare no conflict of interest. The funders had no role in the design of the study; in the collection, analyses, or interpretation of data; in the writing of the manuscript, or in the decision to publish the results.

Abbreviations

The following abbreviations are used in this manuscript:

ABS	Acrylonitrile butadiene styrene
ADD-GSIM	Analysis of distortion distribution-based Gradient SIMilarity
ADD-SSIM	Analysis of distortion distribution-based Structural SIMilarity
CAD	Computer-Aided Design
CISI	Combined Image Similarity Index
CSF	Contrast Sensitivity Function
CSSIM	Color Structural SIMilarity
CVSSI	Contrast and Visual Saliency Similarity-Induced Index
CW-SSIM	Complex Wavelet Structural SIMilarity
DCT	Discrete Cosine Transform
CLAHE	Contrast Limited Adaptive Histogram Equalization
DSS	DCT subband similarity

ESSIM	Edge Strength SIMilarity
FDM	Fused deposition modeling
FR IQA	Full-reference image quality assessment
FSIM	Feature Similarity
GM	Gradient magnitude
GSM	Gradient Similarity
HOG	Histogram of Oriented Gradients
HVS	Human Visual System
IQA	Image quality assessment
ITU	International Telecommunication Union
IW-SSIM	Information Content Weighted Structural SIMilarity
KROCC	Kendall Rank Order Correlation Coefficient
LPBF	Laser powder bed fusion
MCSD	Multiscale Contrast Similarity Deviation
MOS	Mean Opinion Score
MSE	Mean Squared Error
MS-SSIM	Multi-Scale Structural SIMilarity
NDT	Non-destructive testing
NR IQA	No-reference image quality assessment
OCT	Optical coherence tomography
PLCC	Pearson's Linear Correlation Coefficient
QILV	Quality Index based on Local Variance
SROCC	Spearman Rank Order Correlation Coefficient
SR-SIM	Spectral residual-based similarity
SSIM	Structural SIMilarity
PC	Phase congruency
PLA	Polyactic acid
PSNR	Peak Signal-to-Noise Ratio
PVC	Polyvinyl chloride
RVSIM	Riesz transform and Visual contrast sensitivity-based feature SIMilarity
UIQI	Universal Image Quality Index
VQEG	Video Quality Experts Group

References

1. Stephens, B.; Azimi, P.; Orch, Z.E.; Ramos, T. Ultrafine particle emissions from desktop 3D printers. *Atmos. Environ.* **2013**, *79*, 334–339. [CrossRef]
2. Azimi, P.; Zhao, D.; Pouzet, C.; Crain, N.E.; Stephens, B. Emissions of Ultrafine Particles and Volatile Organic Compounds from Commercially Available Desktop Three-Dimensional Printers with Multiple Filaments. *Environ. Sci. Technol.* **2016**, *50*, 1260–1268. [CrossRef] [PubMed]
3. Fastowicz, J.; Grudziński, M.; Tecław, M.; Okarma, K. Objective 3D Printed Surface Quality Assessment Based on Entropy of Depth Maps. *Entropy* **2019**, *21*, 97. [CrossRef]
4. Okarma, K.; Fastowicz, J. Improved quality assessment of colour surfaces for additive manufacturing based on image entropy. *Pattern Anal. Appl.* **2020**, *23*, 1035–1047. [CrossRef]
5. Cheng, Y.; Jafari, M.A. Vision-Based Online Process Control in Manufacturing Applications. *IEEE Trans. Autom. Sci. Eng.* **2008**, *5*, 140–153. [CrossRef]
6. Delli, U.; Chang, S. Automated Process Monitoring in 3D Printing Using Supervised Machine Learning. *Procedia Manuf.* **2018**, *26*, 865–870. [CrossRef]
7. Szkilnyk, G.; Hughes, K.; Surgenor, B. Vision Based Fault Detection of Automated Assembly Equipment. In Proceedings of the ASME/IEEE International Conference on Mechatronic and Embedded Systems and Applications, Parts A and B, Qingdao, China, 15–17 July 2010; ASMEDC: Washington, DC, USA, 2011; Volume 3, pp. 691–697. [CrossRef]
8. Chauhan, V.; Surgenor, B. A Comparative Study of Machine Vision Based Methods for Fault Detection in an Automated Assembly Machine. *Procedia Manuf.* **2015**, *1*, 416–428. [CrossRef]

9. Chauhan, V.; Surgenor, B. Fault detection and classification in automated assembly machines using machine vision. *Int. J. Adv. Manuf. Technol.* **2017**, *90*, 2491–2512. [CrossRef]
10. Fang, T.; Jafari, M.A.; Bakhadyrov, I.; Safari, A.; Danforth, S.; Langrana, N. Online defect detection in layered manufacturing using process signature. In Proceedings of the IEEE International Conference on Systems, Man and Cybernetics, San Diego, CA, USA, 14 October 1998; IEEE: Piscataway, NJ, USA, 1998; Volume 5, pp. 4373–4378. [CrossRef]
11. Fang, T.; Jafari, M.A.; Danforth, S.C.; Safari, A. Signature analysis and defect detection in layered manufacturing of ceramic sensors and actuators. *Mach. Vis. Appl.* **2003**, *15*, 63–75. [CrossRef]
12. Gardner, M.R.; Lewis, A.; Park, J.; McElroy, A.B.; Estrada, A.D.; Fish, S.; Beaman, J.J.; Milner, T.E. In situ process monitoring in selective laser sintering using optical coherence tomography. *Opt. Eng.* **2018**, *57*, 1. [CrossRef]
13. Lane, B.; Moylan, S.; Whitenton, E.P.; Ma, L. Thermographic measurements of the commercial laser powder bed fusion process at NIST. *Rapid Prototyp. J.* **2016**, *22*, 778–787. [CrossRef] [PubMed]
14. Busch, S.F.; Weidenbach, M.; Fey, M.; Schäfer, F.; Probst, T.; Koch, M. Optical Properties of 3D Printable Plastics in the THz Regime and their Application for 3D Printed THz Optics. *J. Infrared Millim. Terahertz Waves* **2014**, *35*, 993–997. [CrossRef]
15. Straub, J. Physical security and cyber security issues and human error prevention for 3D printed objects: detecting the use of an incorrect printing material. In *Dimensional Optical Metrology and Inspection for Practical Applications VI, Proceedings of the SPIE Commercial + Scientific Sensing and Imaging, Anaheim, CA, USA, 9–13 April 2017*; Harding, K.G., Zhang, S., Eds.; SPIE: Bellingham, WA, USA, 2017; Volume 10220, pp. 90–105. [CrossRef]
16. Straub, J. Identifying positioning-based attacks against 3D printed objects and the 3D printing process. In *Pattern Recognition and Tracking XXVIII, Proceedings of the SPIE Defense + Security, Anaheim, CA, USA, 16 June 2017*; Alam, M.S., Ed.; SPIE: Bellingham, WA, USA, 2017; Volume 10203. [CrossRef]
17. Zeltmann, S.E.; Gupta, N.; Tsoutsos, N.G.; Maniatakos, M.; Rajendran, J.; Karri, R. Manufacturing and Security Challenges in 3D Printing. *JOM* **2016**, *68*, 1872–1881. [CrossRef]
18. Holzmond, O.; Li, X. In situ real time defect detection of 3D printed parts. *Addit. Manuf.* **2017**, *17*, 135–142. [CrossRef]
19. Scime, L.; Beuth, J. Anomaly detection and classification in a laser powder bed additive manufacturing process using a trained computer vision algorithm. *Addit. Manuf.* **2018**, *19*, 114–126. [CrossRef]
20. Tourloukis, G.; Stoyanov, S.; Tilford, T.; Bailey, C. Data driven approach to quality assessment of 3D printed electronic products. In Proceedings of the 38th International Spring Seminar on Electronics Technology (ISSE), Eger, Hungary, 6–10 May 2015; IEEE: Piscataway, NJ, USA, 2015; pp. 300–305. [CrossRef]
21. Makagonov, N.G.; Blinova, E.M.; Bezukladnikov, I.I. Development of visual inspection systems for 3D printing. In Proceedings of the 2017 IEEE Conference of Russian Young Researchers in Electrical and Electronic Engineering (EIConRus), St. Petersburg, Russia, 1–3 February 2017; IEEE: Piscataway, NJ, USA, 2017; pp. 1463–1465. [CrossRef]
22. Fok, K.; Cheng, C.; Ganganath, N.; Iu, H.H.; Tse, C.K. An ACO-Based Tool-Path Optimizer for 3-D Printing Applications. *IEEE Trans. Ind. Inform.* **2019**, *15*, 2277–2287. [CrossRef]
23. Sitthi-Amorn, P.; Ramos, J.E.; Wangy, Y.; Kwan, J.; Lan, J.; Wang, W.; Matusik, W. MultiFab: A Machine Vision Assisted Platform for Multi-material 3D Printing. *ACM Trans. Graph.* **2015**, *34*, 1–11. [CrossRef]
24. Straub, J. Initial Work on the Characterization of Additive Manufacturing (3D Printing) Using Software Image Analysis. *Machines* **2015**, *3*, 55–71. [CrossRef]
25. Straub, J. An approach to detecting deliberately introduced defects and micro-defects in 3D printed objects. In *Pattern Recognition and Tracking XXVIII, Proceedings of the SPIE Defense + Security, Anaheim, CA, USA, 16 June 2017*; Alam, M.S., Ed.; SPIE:Bellingham, WA, USA, 2017; Volume 10203. [CrossRef]
26. Bi, Q.; Wang, M.; Lai, M.; Lin, J.; Zhang, J.; Liu, X. Automatic surface inspection for S-PVC using a composite vision-based method. *Appl. Opt.* **2020**, *59*, 1008. [CrossRef]
27. Zhai, G.; Min, X. Perceptual image quality assessment: A survey. *Sci. China Inf. Sci.* **2020**, *63*. [CrossRef]
28. Wang, Z.; Bovik, A. A universal image quality index. *IEEE Signal Process. Lett.* **2002**, *9*, 81–84. [CrossRef]
29. Wang, Z.; Bovik, A.; Sheikh, H.; Simoncelli, E. Image Quality Assessment: From Error Visibility to Structural Similarity. *IEEE Trans. Image Process.* **2004**, *13*, 600–612. [CrossRef] [PubMed]

30. Wang, Z.; Simoncelli, E.; Bovik, A. Multiscale structural similarity for image quality assessment. In Proceedings of the 37th Asilomar Conference on Signals, Systems and Computers, Pacific Grove, CA, USA, 9–12 November 2003; IEEE: Piscataway, NJ, USA, 2003; pp. 1398–1402. [CrossRef]
31. Aja-Fernandez, S.; Estepar, R.S.J.; Alberola-Lopez, C.; Westin, C.F. Image Quality Assessment based on Local Variance. In Proceedings of the 2006 International Conference of the IEEE Engineering in Medicine and Biology Society, New York, NY, USA, 30 August–3 September 2006; IEEE: Piscataway, NJ, USA, 2006. [CrossRef]
32. Sampat, M.; Wang, Z.; Gupta, S.; Bovik, A.; Markey, M. Complex Wavelet Structural Similarity: A New Image Similarity Index. *IEEE Trans. Image Process.* **2009**, *18*, 2385–2401. [CrossRef] [PubMed]
33. Wang, Z.; Li, Q. Information Content Weighting for Perceptual Image Quality Assessment. *IEEE Trans. Image Process.* **2011**, *20*, 1185–1198. [CrossRef] [PubMed]
34. Zhang, L.; Zhang, L.; Mou, X.; Zhang, D. FSIM: A Feature Similarity Index for Image Quality Assessment. *IEEE Trans. Image Process.* **2011**, *20*, 2378–2386. [CrossRef] [PubMed]
35. Liu, A.; Lin, W.; Narwaria, M. Image Quality Assessment Based on Gradient Similarity. *IEEE Trans. Image Process.* **2012**, *21*, 1500–1512. [CrossRef]
36. Zhang, L.; Li, H. SR-SIM: A fast and high performance IQA index based on spectral residual. In Proceedings of the 2012 19th IEEE International Conference on Image Processing (ICIP), Orlando, FL, USA, 30 September–3 October 2012; pp. 1473–1476. [CrossRef]
37. Zhang, X.; Feng, X.; Wang, W.; Xue, W. Edge Strength Similarity for Image Quality Assessment. *IEEE Signal Process. Lett.* **2013**, *20*, 319–322. [CrossRef]
38. Balanov, A.; Schwartz, A.; Moshe, Y.; Peleg, N. Image quality assessment based on DCT subband similarity. In Proceedings of the 2015 IEEE International Conference on Image Processing (ICIP), Quebec City, QC, Canada, 27–30 September 2015; IEEE: Piscataway, NJ, USA, 2015. [CrossRef]
39. Wang, T.; Zhang, L.; Jia, H.; Li, B.; Shu, H. Multiscale contrast similarity deviation: An effective and efficient index for perceptual image quality assessment. *Signal Process. Image Commun.* **2016**, *45*, 1–9. [CrossRef]
40. Gu, K.; Wang, S.; Zhai, G.; Lin, W.; Yang, X.; Zhang, W. Analysis of Distortion Distribution for Pooling in Image Quality Prediction. *IEEE Trans. Broadcast.* **2016**, *62*, 446–456. [CrossRef]
41. Ponomarenko, M.; Egiazarian, K.; Lukin, V.; Abramova, V. Structural Similarity Index with Predictability of Image Blocks. In Proceedings of the 2018 IEEE 17th International Conference on Mathematical Methods in Electromagnetic Theory (MMET), Kiev, Ukraine, 2–5 July 2018; IEEE: Piscataway, NJ, USA, 2018. [CrossRef]
42. Jia, H.; Zhang, L.; Wang, T. Contrast and Visual Saliency Similarity-Induced Index for Assessing Image Quality. *IEEE Access* **2018**, *6*, 65885–65893. [CrossRef]
43. Yang, G.; Li, D.; Lu, F.; Liao, Y.; Yang, W. RVSIM: A feature similarity method for full-reference image quality assessment. *EURASIP J. Image Video Process.* **2018**, *2018*, 6. [CrossRef]
44. Ieremeiev, O.; Lukin, V.; Okarma, K.; Egiazarian, K. Full-Reference Quality Metric Based on Neural Network to Assess the Visual Quality of Remote Sensing Images. *Remote Sens.* **2020**, *12*, 2349. [CrossRef]
45. Okarma, K. Combined Full-Reference Image Quality Metric Linearly Correlated with Subjective Assessment. In *Artificial Intelligence and Soft Computing, Proceedings of the 10th International Conference Proceedings, ICAISC 2010, Zakopane, Poland, 13–17 June 2010*; Rutkowski, L., Scherer, R., Tadeusiewicz, R., Zadeh, L.A., Zurada, J.M., Eds.; Springer: Berlin/Heidelberg, Germany, 2010; Volume 6113, pp. 539–546. [CrossRef]
46. Sheikh, H.; Bovik, A. Image information and visual quality. *IEEE Trans. Image Process.* **2006**, *15*, 430–444 [CrossRef] [PubMed]
47. Mansouri, A.; Aznaveh, A.M.; Torkamani-Azar, F.; Jahanshahi, J.A. Image quality assessment using the singular value decomposition theorem. *Opt. Rev.* **2009**, *16*, 49–53. [CrossRef]
48. Okarma, K. Combined image similarity index. *Opt. Rev.* **2012**, *19*, 349–354. [CrossRef]
49. Athar, S.; Wang, Z. A Comprehensive Performance Evaluation of Image Quality Assessment Algorithms. *IEEE Access* **2019**, *7*, 140030–140070. [CrossRef]
50. Oszust, M. A Regression-Based Family of Measures for Full-Reference Image Quality Assessment. *Meas. Sci. Rev.* **2016**, *16*, 316–325. [CrossRef]
51. Liu, T.J.; Lin, W.; Kuo, C.C.J. Image Quality Assessment Using Multi-Method Fusion. *IEEE Trans. Image Process.* **2013**, *22*, 1793–1807. [CrossRef]

52. Lukin, V.V.; Ponomarenko, N.N.; Ieremeiev, O.I.; Egiazarian, K.O.; Astola, J. Combining full-reference image visual quality metrics by neural network. In *Human Vision and Electronic Imaging XX, Proceedings of the SPIE/IS&T Electronic Imaging, San Francisco, CA, USA, 10–11 February 2015*; Rogowitz, B.E., Pappas, T.N., de Ridder, H., Eds.; SPIE: Bellingham, WA, USA, 2015; Volume 9394. [CrossRef]
53. Oszust, M. Full-Reference Image Quality Assessment with Linear Combination of Genetically Selected Quality Measures. *PLoS ONE* **2016**, *11*, e0158333. [CrossRef]
54. Okarma, K.; Fastowicz, J.; Tecław, M. Application of Structural Similarity Based Metrics for Quality Assessment of 3D Prints. In *Computer Vision and Graphics, Proceedings of the International Conference, ICCVG 2016, Warsaw, Poland, 19–21 September 2016*; Chmielewski, L.J., Datta, A., Kozera, R., Wojciechowski, K., Eds.; Springer International Publishing: Cham, Switzerland, 2016; Volume 9972, pp. 244–252. [CrossRef]
55. Okarma, K.; Fastowicz, J. Adaptation of Full-Reference Image Quality Assessment Methods for Automatic Visual Evaluation of the Surface Quality of 3D Prints. *Elektron. Elektrotechnika* **2019**, *25*, 57–62. [CrossRef]
56. Fastowicz, J.; Okarma, K. Fast quality assessment of 3D printed surfaces based on structural similarity of image regions. In Proceedings of the 2018 International Interdisciplinary PhD Workshop (IIPhDW), Świnoujście, Poland, 9–12 May 2018. [CrossRef]
57. International Telecommunication Union. *Recommendation ITU-R BT.601-7—Studio Encoding Parameters of Digital Television for Standard 4:3 and Wide-Screen 16:9 Aspect Ratios*; International Telecommunication Union: Geneva, Switzerland, 2011.
58. Fastowicz, J.; Lech, P.; Okarma, K. Combined Metrics for Quality Assessment of 3D Printed Surfaces for Aesthetic Purposes: Towards Higher Accordance with Subjective Evaluations. In *Computational Science—ICCS 2020, Proceedings of the 20th International Conference Proceedings, Amsterdam, The Netherlands, 3–5 June 2020*; Krzhizhanovskaya, V.V., Závodszky, G., Lees, M.H., Dongarra, J.J., Sloot, P.M.A., Brissos, S., Teixeira, J., Eds.; Springer International Publishing: Cham, Switzerland, 2020; Volume 12143, pp. 326–339. [CrossRef]
59. Kim, H.; Lin, Y.; Tseng, T.L.B. A review on quality control in additive manufacturing. *Rapid Prototyp. J.* **2018**, *24*, 645–669. [CrossRef]
60. Wu, H.C.; Chen, T.C.T. Quality control issues in 3D-printing manufacturing: A review. *Rapid Prototyp. J.* **2018**, *24*, 607–614. [CrossRef]

© 2020 by the authors. Licensee MDPI, Basel, Switzerland. This article is an open access article distributed under the terms and conditions of the Creative Commons Attribution (CC BY) license (http://creativecommons.org/licenses/by/4.0/).

MDPI
St. Alban-Anlage 66
4052 Basel
Switzerland
Tel. +41 61 683 77 34
Fax +41 61 302 89 18
www.mdpi.com

Applied Sciences Editorial Office
E-mail: applsci@mdpi.com
www.mdpi.com/journal/applsci

www.ingramcontent.com/pod-product-compliance
Lightning Source LLC
LaVergne TN
LVHW070658100526
838202LV00013B/992